Outsourcing Energy Management

To Derrick Gerald Fawkes
1931 to 2004

Outsourcing Energy Management

Saving Energy and Carbon Through Partnering

STEVEN FAWKES

GOWER

Published by
Gower Publishing Limited
Gower House
Croft Road
Aldershot
Hampshire GU11 3HR
England

Ashgate Publishing Company
Suite 420
101 Cherry Street
Burlington, VT 05401-4405
USA

British Library Cataloguing in Publication Data
Fawkes, Steven
Outsourcing energy management : saving energy and carbon through partnering
1. power resources – Management 2. Contracting out
I. Title
658.2'6

ISBN-13: 9780566087127

Library of Congress Cataloging-in-Publication Data
Fawkes, Steven.
Outsourcing energy management: saving energy and carbon through partnering / by Steven Fawkes.
p. cm.
Includes bibliographical references and index.
ISBN 978-0-566-08712-7
1. Energy conservation. 2. Contracting out. I. Title.
TJ163.3.F39 2007
658.2'6--dc22

2006033549

Printed and bound in Great Britain by TJ International Ltd, Padstow, Cornwall.

Contents

List of Figures

List of Tables

List of Boxes

List of Abbreviations

AC	alternating current
aM&T	automatic M&T
AMR	automated meter reading
BAT	best available technique
BEMS	building energy management system
BIPV	building integrated photo-voltaics
BMS	building management system
BOE	barrels of oil equivalent
BOO	build, own and operate
CCL	Climate Change Levy
CDM/JI	Clean Development Mechanisms/Joint Initiatives
CEM	contract energy management
CH4	methane – powerful greenhouse gas
CHP	combined heat and power
CHPQA	Combined Heat and Power Quality Assurance scheme
CI	continuous improvement
CO2	carbon dioxide
CoP	coefficient of performance
CPI	consumer price index
CSR	corporate social responsibility
DC	direct current
DCF	discounted cash flow
DTI	Department of Trade and Industry
DUOS	Distribution Use of Systems charge

EA	Environment Agency
EDI	electronic data interchange
ESCO	energy service company
EPC	energy performance contract
ESPC	energy service provider company
ESTA	Energy Systems Trade Association
EU ETS	EU Emissions Trading Scheme
FSU	former Soviet Union
GHG	greenhouse gases
GSHP	ground source heat pump
HPHW	high pressure hot water
HR	human resources
HV	high voltage
IFC	International Finance Corporation
IGA	investment grade audit
IPCC	Intergovernmental Panel on Climate Change
IPMV	International Protocol on Measurement and Verification
IPPC	Integrated Pollution Prevention and Control
IRR	internal rate of return
KPI	key performance indicator
kW	unit of power: 1000 watts
kWh	unit of energy: 1000 watt-hours
LDs	liquidated damages
LED	light emitting diode
LV	low voltage
M&T	monitoring and targeting
M&V	measurement and verification
MPHW	medium pressure hot water

MW	unit of power: 1 000 000 watts
MWh	unit of energy: 1 000 000 watt-hours
NAP	National Allocation Plan – allocation of carbon dioxide emissions under EU ETS
NPI	normalised performance indicator
NPV	net present value
O&M	operations and maintenance
OFR	operating and financial review
OPEC	Organization of the Petroleum Exporting Countries
PFI	private finance initiative
PPP	public private partnership
PV	photovoltaic
RFP	request for proposals
RFQ	request for qualification
ROC	renewable obligation certificate
ROCE	return on capital employed
RPI	retail price index
SCADA	sequencing control and data acquisition
SEC	specific energy consumption – energy consumption per unit of output
SLA	service level agreement
SME	small or medium enterprise (< 250 employees)
SPV	special purpose vehicle
THC	thermohaline cycle
TUOS	Transmission Use of Systems charge
TUPE	Transfer of Undertakings Protected Employment
UNEP	United Nations Environment Programme
UNFCC	United Nations Framework on Climate Change

USAID	United States Agency for International Development
VSD	variable speed drive
WMO	World Meteorological Organisation
WWF	World Wildlife Fund

Acknowledgements

I would like to acknowledge all those who have contributed to my work and life and without whom this book would not have been possible, particularly the following: my parents, my sister, Noel Billing, Keith Jacques, Rex Wellard, Kym Coe, Ron Smothermon, Michael Roberts, Don Carey, Tim Lines, Mark Velody, William Rae, Mike Dean, Henry Raschen, David Anderson, Andrew Bainbridge, Mark Hawker, Brian Mackle, Julius Brinkworth, Susan Babb, Julie Leff, Paula Bright, Morgan Lobb, Chris Lamb, Patrick Bastien, Paul Turner, John Oliver, Walt Patterson, Patrick Collins, Peter and Francis Morimura, Catalin Dragostin, Scott White, Ray and Connie White, William Fortescue, Andrew MacDonald, and the npower Solutions team. Most of all I would like to acknowledge my wife Kathy for all her encouragement and support. As usual any errors or omissions are the responsibility of the author.

Preface

The most technologically progressive century in history ended with fuel technologies not fundamentally different from those when it began.

John Kay, *Financial Times*,
26 July 2005

In 10–15 years' time we will be dependent on very expensive oil and gas largely from the former Soviet Union and the Middle East – hardly reliable suppliers. Oil production will peak in the next few years, gas after another decade. I write as a retired hydrocarbon explorer with nearly 40 years' experience and considerable success (including Brent, which I named) with Shell and Enterprise Oil. There is not that much oil out there to be found, despite what the big fuel companies (and governments) say. How many are able to replace their annual production by their own successful exploration efforts? As far as I know – none. The future is bleak, even catastrophic.

Dr Myles Bowen, OBE, letter to the *Sunday Times*,
25 September 2005

This book started life as a proposal to write a book on energy management techniques and the available technologies to improve energy efficiency for industrial, commercial and public sector organisations. Most books on the subject were written in the late 1970s or early 1980s, at a time of high energy prices and increased interest in energy management. In the 1990s real energy prices declined, partially under the influence of the market liberalisation and privatisations led by the UK, and partly due to abundant supply of oil on the world market. Today the picture looks very different again. Oil, gas and electricity prices have risen dramatically in the last few years, and the impacts of oil peaking and concerns about climate change mean that it is likely that they will stay high, and increase in the coming years. The problem of oil peaking, reaching a maximum level of global oil production and then declining, is moving from being regarded as a fringe concern to centre stage. The effect of carbon emissions from fossil fuel combustion on the global environment, global warming, has also been recognised and accepted by almost all serious analysts. Even if there was no direct cause and effect, the impact of Hurricanes

Katrina and Rita on the US Gulf of Mexico coast in late 2005 led to even greater recognition of both global warming and the issues surrounding fossil fuel supply.

It is clear that over the next 20 to 50 years we must switch our energy system to one that is not based on fossil fuels. After some false starts we are entering the end of the fossil fuel era and beginning to assemble the elements of a cleaner, safer energy system in which improved energy efficiency and renewable energy have large parts to play. The exact mix of energy sources in the future is still unclear and is a matter for much debate: renewables versus nuclear, centralised versus decentralised. Improving energy efficiency has to be a central part of any policy. Arthur C. Clarke, the science fiction writer, once said that we tend to over-estimate how much progress we will make in the short-term future and under-estimate how much progress we will make in the long-term future. I believe this will be true of the energy situation and the energy world in 50 years will probably contain some big surprises.

There has been no change in the basic processes of energy demand management since they were first formalised in the early to mid-1980s. Much of the technology to improve energy efficiency also remains essentially the same, although the revolution of micro-electronics, computers and communications has, as in other spheres, greatly enhanced capabilities and reduced the real price of monitoring and control equipment as well as data collection and processing. Some new technologies are emerging, such as decentralised, embedded generation from small, high-efficiency combined heat and power (CHP) units, small-scale renewable energy systems; and in the area of space heating, ground source heat pumps. Some emerging technologies such as light emitting diode (LED) lighting, building integrated photovoltaics (BIPV) and fuel cell CHP show great potential to achieve step changes in efficiency and emissions reductions although they have not yet reached the point of satisfying normal economic criteria.

Since the mid-1980s most organisations have lost much of their capacity and capability to manage energy. This has been due to two factors: low energy prices meant the emphasis switched to procurement, that is buying energy at the lowest price and away from physically managing its use; and the general downsizing of organisations seen throughout the 1990s and early years of this decade. Reduced capacity was the motivator for writing a book on energy management, but in considering and discussing the issues facing organisations today it became clear that outsourcing of energy services was a more interesting topic than 'straightforward' energy management. Outsourcing in general is a growth trend and energy outsourcing is also growing as organisations grapple with the twin problems of higher energy prices and responding to the carbon agenda with reduced energy management capacity. Most of the outsourcing of energy management activities has been partial as the supply of total, integrated models has not been generally available. Much of the available literature on the subject is focused on the American approach of energy savings performance contracting and although this is a very successful model, it is not the only one and nor does it address the total problem. At the same time most of the UK experience has been in contract energy management, usually driven more by operations and maintenance considerations rather than energy efficiency. I wanted to examine the rationale and process of outsourcing energy services as well as introduce new models, models driven by what customers want and need rather than by what is currently available.

I have worked in energy management for over 25 years and most of that time has been spent providing energy services, a mixture of pure consulting and investment-led services. In that time I have worked in a wide range of organisations on all aspects of energy management, including corporate strategy, government policy, monitoring and targeting, developing and implementing investment programmes and motivation and training. I have also been involved in creating and developing three investment-based energy services companies as well as several consultancy companies. The investment-led companies developed and implemented novel approaches to energy outsourcing and I wanted to share some of the experience gained. The new challenges of global sustainability, specifically climate change and oil peaking, require new solutions, and this book is part of working out these solutions.

The book is designed for managers of organisations at all levels, from CEOs considering outsourcing of energy services to those managers who have to design, implement and manage projects. It is designed to guide thinking on the subject and help you shortcut some of the problems encountered elsewhere. It is essentially a 'how to do it' book, although the first section deals with 'why you should do it'. It is also designed to be useful to professionals working for energy service companies and will be of interest to students of energy matters. I trust that it is a useful contribution to resolving the current challenges and exploiting the opportunities.

Steven Fawkes

PART 1

The Importance of Energy

INTRODUCTION

> ***Energy*** *is a quantifiable state function of every physical system. Energy allows one to predict how much work a physical system could be made to do, or how much heat it can exchange.*
>
> Brainy Encyclopaedia.com

This introductory section sets out the reasons why managers of companies and other organisations, and, in fact, why all of us should have considerable concern about the issues of energy supply and demand. It outlines the UK and global energy situations without attempting to be a technical work (as with the rest of this book it is designed to be read by non-energy professionals), because understanding the context of the energy issue is important both in order to guide and motivate informed decision making. Sometimes, to the annoyance of energy professionals, energy demand decisions that ultimately drive supply decisions are taken by many diverse groups and individuals and are not driven by energy issues, but rather by many other technological, economic, social and emotional factors. For instance:

- Do we invest in a more efficient boiler when we have three other projects competing for scarce capital that will increase sales or improve quality?
- We need a new building quickly, and energy is only a small proportion of total costs; why should we bother about its energy efficiency?
- Does reducing carbon emissions really matter, compared to making a profit?

This book is based firmly in the reality of the knowledge that these kinds of questions that have to be answered every day in organisations. I believe they can only be answered optimally with a good understanding of the UK and global energy context.

The large-scale use of energy, fuel and electricity is literally the lifeblood of modern society. Without it we cannot manufacture goods, we cannot travel or transport goods, we cannot grow food in the quantities and at the prices we are used to, and ultimately we cannot survive in environments outside a narrow band of temperatures.

Society is facing a number of challenges over the coming decades that include international terrorism, the proliferation of nuclear weapons and how to bring the developing world up to a standard of living comparable to that of developed countries. All of these have direct links to energy supply and demand. They are also all taking place in the context of what may turn out to be the largest problem – global climate change caused by human activities. We cannot develop the world such that a global population of ten billion has the benefit of the energy use that the developed world currently enjoys using the existing fossil fuel-based energy systems. Fossil fuels would quickly be depleted and the effects on the global climate would be far larger than anything seen so far, possibly catastrophic. Even without considering the development issue, the continued use of fossil fuels is ultimately unsustainable due to their finite nature.

The section is divided into three chapters dealing with the three main concerns that make energy an important issue, irrespective of the price of energy.

First, we deal with the issue of the finite nature of fossil fuels, the lifeblood of our industrial and post-industrial society. This is considered by most to be a long-term problem, but as we will see there are those analysts who consider it very much a short-term problem. As the source of most of our fossil fuel is from potentially unstable regions of the world, particularly the Middle East, and there is the ever-present threat of terrorism on a large scale, fossil fuel supplies could be constrained at any time by geopolitical events, irrespective of the physical availability of oil and gas in the ground.

Second, we deal with the threat of climate change, a threat that is driving policy in many countries. Climate change can be considered a medium-term concern that demands attention now.

Third, we deal with the existing, and foreseeable, legislative and regulatory pressures that have been added to by the first two concerns. These can be considered short-term concerns because the laws and regulations discussed are already in existence or are coming soon. Non-compliance is not an option for any organisation.

In addition to these issues, organisations of all types have been facing greatly increased energy prices and energy price volatility. Since 2002 energy prices for some customers in the industrial market have risen between 50 and 100 per cent. Short-term spikes in gas prices have been particularly extreme. This sudden and extreme price increase has put energy and the control of energy costs firmly back on the agenda for directors and senior managers of most organisations. For some energy intensive manufacturers, the increased prices are threatening competitiveness and even their survival with a growing number of sites being closed or suspending operations due to high energy prices. Several companies in the UK linked the rise in energy prices to reduced profits in their annual general meetings. This is the immediate problem facing managers of all organisations today.

This confluence of problems, combined with increasing and more volatile energy prices, the ageing energy infrastructure within many organisations and the decline in capacity to manage energy which occurred during the period of low and stable energy prices, means that there is a massive need for a re-allocation of resources in favour of improving energy efficiency and implementing cleaner, low carbon energy solutions. This re-allocation of resources will have to come from all energy users, energy suppliers and financial institutions.

Energy users can have three responses to the energy problems discussed above:

1. do nothing;
2. implement programmes focused purely on cost saving;
3. view the problems as a strategic opportunity and respond accordingly.

The scale and nature of the problems suggests that doing nothing is not really an option for credible companies interested in a long-term future. If energy prices do not force such companies out of business surely customer demands and increasing regulations will do so. For many organisations the proper response will be to implement a programme focused on energy cost saving, and this book outlines methods and techniques that will allow maximum savings to be achieved. For some organisations, and the number is growing, the response to the energy and environmental problems will be to see them as a strategic opportunity to differentiate themselves in a crowded marketplace. Such organisations are likely to implement programmes that go beyond pure cost saving and move into carbon management and projects such as renewable energy, which may not yet be fully economic. They believe that the indirect benefits such as corporate social responsibility reputation and improved risk management elements of these projects outweigh simple project economics.

For organisations choosing either a cost-saving response or a strategic opportunity response, the problems will include accessing sufficient capital, human and technical resources. This is where outsourcing of some or all of the energy management process can be a useful tool.

CHAPTER 1

Oil Peaking and the Decline of Fossil Fuels

The fifth revolution will come when we have spent the stores of coal and oil that have been accumulating in the earth during hundreds of millions of years... It is to be hoped that before then other sources of energy will have been developed... Whether a convenient substitute for the present fuels is found or not, there can be no doubt that there will have to be a great change in ways of life. This change may justly be called a revolution, but it differs from all the preceding ones in that there is no likelihood of its leading to increases of population, but even perhaps to the reverse.

Sir Charles Galton Darwin, 1952

Oil has become more and more necessary to everyone. At first it was used for medicine, for lamps, and for lubrication. Now, if the supply of oil were cut off, our manner of living would change completely until something to take its place was found.

Maud and Miska Petersham, 1935 (when the world used 3.5 million barrels of oil each day compared to the 80 million barrels of oil used each day in 2004)

Energy will be one of the defining issues of this century, and one thing is clear: the era of easy oil is over.

Chevron website, 2006 (www.willyoujoinus.com/vision/)

[T]he 'hidden cost' of a gallon of imported oil is now between $5.28 and $5.55. When added to the nominal pump price of unleaded gasoline, this yields a total of between $8.26 and $8.53 per gallon… the total 'hidden cost' of our oil import dependence now totals between $524.9 billion and $532.6 billion annually – over 2.5 times what we have spent on the war in Iraq.

National Defense Council Foundation USA website, 2005 (www.ndcf.org/reports/AlaskasStrandedGas.pdf)

INTRODUCTION

Before considering the supply of fossil fuels (oil, gas and coal), we need to remind ourselves just how dependent modern society is on fossil fuel usage. Industrial society's use of fossil fuels began in the 1800s with the use of coal. Starting at the beginning of the twentieth century there was a shift to oil rather than coal, initially prompted by naval use of oil for military ships as it enabled longer range and reduced labour forces. This geopolitical motivation quickly led to a rapid expansion of the oil industry up to and after World War II, with polices predicated on cheap oil, mostly supplied by the Middle East. Currently fossil fuels provide 88 per cent of the world energy consumption, with nuclear and hydroelectric accounting for 6 per cent each, with solar and biomass accounting for less than 1 per cent, while in the UK coal, oil and natural gas account for 89 per cent of total energy use (DTI, 2006).

There has been strong growth in primary energy consumption since 1979. In 2004 global primary energy consumption recorded the strongest incremental growth ever, rising by 4.3 per cent. Growth was above the ten-year average in all regions and for all fuels (BP, 2005). In the UK primary energy consumption grew by 13 per cent between 1970 and 2001, and by 11 per cent between 1990 and 2001, and growth has continued since then.

Global society is now totally dependent on the large-scale industrial use of fuels and electricity and the global energy business is huge. The energy industry is the world's largest industrial activity with revenues of $8 to $9 trillion (Simmons, 2006). In the UK total expenditure on energy (excluding transport) in 2004 was £23 billion, a figure which has probably increased by 50 per cent since then due to energy price increases (DTI, 2006).

Despite the enormous scale of global energy use, the majority of the world's population of six billion people do not yet benefit from industrial energy use. Ensuring that they, and the expected future population of ten billion, do so but without environmental catastrophe or armed conflict is one of the major challenges facing humanity in for the twenty-first century.

Given the dependence of global society on fossil fuels, and particularly oil, a fundamental issue is how long the available sources of oil will last.

HUBBERT'S CURVE

There will always be debate about the amount of oil and gas resources left on the planet but there are two inescapable facts about oil and all other fossil fuels: there is a limited supply and if demand continues to increase, at some point demand will outstrip production.

Box 1.1 Fossil fuels: quantification and definitions

While oil is the most ubiquitous form of fossil fuel, coal and natural gas can be similarly categorised. All three forms of fossil fuel can be quantified in terms of barrels of oil equivalent (BOE). One BOE equals 6000 cubic feet of natural gas. For coal, one BOE equals 0.2 (hard coal) to 0.4 (light coal) tonnes. Current estimates of known reserves for these three forms of hydrocarbons are:

Oil: 1.213 trillion barrels
Gas: 885 billion BOE
Coal: 3.2 trillion BOE

Between these fossil fuels the world consumes 61.7 billion BOE per year, which is 88 per cent of the total energy consumption of 70.4 billion BOE per year.

Source: Guinness Atkinson Funds (2004). Investment Research Series. *The Future of Energy*. Woodland Hills, CA

The best-known tool for predicting the remaining lifetime of fossil fuel reserves is the Hubbert curve. Dr M. King Hubbert was a geologist who worked for Shell and, in 1956, using a simple bell-shaped or logistical curve to look at fossil fuel use and resource life for the United States, predicted that US oil production would peak in 1970. The Hubbert curve was widely criticised but it proved to be remarkably effective. US oil production peaked in 1970–1971, although it was several years before this was acknowledged and Hubbert's work was fully recognised.

The Hubbert curve has subsequently been applied to global oil supplies, and although still controversial, the idea of 'oil peaking' being a major problem in the short term is becoming widely accepted. It is important to be clear that oil peaking is not about 'running out' of oil. Peak output is the point at which maximum volume is produced, after which the peak production declines. As new fields are found these fields will be exploited as fast as possible to meet increasing demand. Peak production occurs when half the reserves have been used and therefore even in the pessimistic scenarios oil will be expected to be available for another 50 to 75 years; the questions are 'how easily can it be found and extracted?' and 'at what price?'

Because oil prices have been relatively high for the past decade, oil companies have conducted extensive exploration over that period, but the results have been disappointing. Annual oil discoveries have been less than annual consumption since 1982 (Aleklett and Campbell, 2002) and it is clear that we are moving from a long period in which additions to reserves were much greater than consumption, to an era in which annual additions are falling increasingly short of annual consumption. This is one of a number of trends that suggest the world is fast approaching the inevitable peaking of conventional world oil production.

So what are the current predictions for when the peak production of oil will occur? The answer is surprisingly close, with a number of experts projecting that peak world production of oil could occur around 2010. It may have already happened by the time you read this book. The table below summarises the range of opinions.

Box 1.2 Predictions of world oil production peaking

2006–2007: Bakhitari, A.M.S. – Iranian oil executive

2007–2009: Simmons, M.R. – Investment banker

After 2007: Skrebowski, G. – Petroleum journal editor

Before 2009: Deffeyes, K.S. – Oil company geologist (retired)

Before 2010: Goodstein, D. – Vice provost, Cal Tech

Around 2010: Campbell, C.J. – Oil company geologist (retired)

After 2010: World Energy Council – Non-governmental organisation

2010–2020: Laherree, J – Oil company geologist (retired)

2016: EIA – United States Department of Energy

After 2020: CERA – Energy consultants

2025 or after: Shell – Major oil company

No visible peak: Lynch, M.C. – Energy economist

Source: Hirsch, R.L., Bezdek, R. and Wending, R. (2005). *Peaking of World Oil Production: Impacts, Mitigation & Risk Management*

One interesting thing about the Hubbert curve is that the year of peak oil production is not very sensitive to the amount of total reserves assumed.

Doubling the estimate of total reserves only adds about 10 years to the peak production year (Leggett, 2005). This illustrates that even if the optimists in the oil industry are right and the oil companies do find new major discoveries, the year of peak production will not be put off very long.

At the same time as oil production is peaking, oil demand is climbing rapidly, particularly in Asia and especially in India and China. In recent years one half of the growth in demand has come from Asia. Looking forward, Asia is likely to continue to be the engine of global oil demand growth. Over the last 17 years Asian oil demand has grown by 10 million barrels daily, and it is hard to see how it will not grow by at least this amount in the next 17 years. In fact, growth is likely to accelerate as Asia goes through the energy intensive state of economic development.

Box 1.3 Increasing Asian oil demand

To illustrate this growth in Asian oil demand it is simply necessary to examine the potential for growth in the Chinese automobile industry. At present, domestic Chinese automobile demand is easing dramatically towards two million vehicles per year. There exists a remarkable parallel between US automobile demand in 1910 and Chinese automobile demand today.

China is producing one car for every 600 people, which equates to the US automobile penetration in 1910. In the years 1910 to 1920, US car production jumped 10-fold to one car per 60 Americans. If China's car production followed a similar trajectory, it would equate to 21 million additional cars in a decade's time. This would mean that China would be approaching the same 150 million automobiles owned by Americans. Given that US gasoline consumption is nine million barrels per day, it is easy to see how China's consumption could rise dramatically from its current level.

Source: Guinness Atkinson Funds (2004). Investment Research Series. *The Future of Energy.* Woodland Hills, CA

So as oil production peaks and then reduces, oil demand will be increasing. Basic supply and demand economics suggests that there is only one result of this situation: increasing prices.

Skrebowski (2005) puts forward a scenario showing supply shortfall and the oil price needed to reduce demand, which is shown below:

2007 2.8 million b/d and $70/barrel

2008 4.8 million b/d and $80/barrel

2009 6.7 million b/d and $90/barrel

2010 8.9 million b/d and $101/barrel

Skrebowski lists the factors that will delay or advance the onset of peak oil.

Factors that will delay the onset of peak oil include:

- economic recession/slowdown;
- demand destruction via high prices;
- systems over performing;
- peace in Iraq;
- Middle East opening up to investment.

Factors that will advance peak oil include:

- slippage in projects to develop new fields;
- increasing taxes (happening);
- accelerating decline in production (happening);
- upheaval in major producers (Iraq, Nigeria, Venezuela already happened);
- accelerating demand growth (China and India happening);
- system breakdowns, wars and revolutions.

Prices during 2006 of over $70 per barrel suggest that this forecast made in 2005 may be accurate.

In the past it has been argued that increasing prices will lead to increased exploration and ultimately increased production. The Hubbert curve, however, shows that this does not happen. Whatever price oil reaches, we will be facing year-on-year declines in production, quite probably within this decade.

Thirty-eight countries can be classified as major oil producers, eleven of which are OPEC (Organization of Petroleum Exporting Countries) members. Collectively these countries are sitting on proven reserves of 902 billion barrels of oil (75 per cent of the total proven reserves). Many OPEC member countries are potentially unstable, and the largest producer is Saudi Arabia, which has many internal strains and has been the target of terrorism, including attacks on oil facilities. Any disruption to oil production in Saudi Arabia would lead to rapid and dramatic price increases and supply disruptions, similar to those experienced after the Yom Kippur war of 1973 and the Iranian revolution of 1979.

CHAPTER 2

Climate Change

Climate change is the most severe problem that we are facing today – more serious even than the threat of terrorism.

Sir David King, UK Government's Chief Scientific Adviser
9 January 2004, *Science*

Climate change is an issue for finance and economic ministries as much as for energy and environmental ones.

Rt Hon Gordon Brown MP, UK Chancellor of the Exchequer
15 March 2005, speech at the Energy and Environment Ministerial Roundtable

Based on our understanding of the range of uncertainty around the scientific views, we've come to the judgement that to avoid serious impact upon societies or the environment, it is necessary to stabilise atmospheric concentrations of greenhouse gases at around 500 to 550 parts per million. Such a shift to a significantly lower carbon economy would require the removal by 2050 of a significant volume of carbon emissions. There is no single solution – no magic bullet. But on the basis of practical steps, using technology which is either available or which may be within reach, stabilisation on that timescale does seem to be an attainable goal.

Lord Browne, Chairman, BP, 25 November 2003

What is now plain is that the emission of greenhouse gases, associated with industrialisation and strong economic growth from a world population that has increased six fold in 200 years, is causing global warming at a rate that began as significant, has become alarming and is simply unsustainable in the long-term. And by long-term I do not mean centuries ahead, I mean within the lifetime of my children certainly; and possibly within my own. And by unsustainable, I do not mean a phenomenon causing problems of adjustment. I mean a challenge so far-reaching in its impact and irreversible in its destructive power, that its alters radically human existence.

Rt Hon Tony Blair, Prime Minister, 14 September 2004

INTRODUCTION

Many governments around the world and their advisers have accepted that climate change is one of the major problems facing the world today. While the science of human-induced global warming may remain uncertain, the issue of what actions to take in the face of that uncertainty remains, but uncertainty cannot be an excuse for taking no action. Rather it implies acting carefully and taking out insurance against the worst risks.

As well as the fundamental supply-and-demand factors that threaten to drive up energy prices (discussed in Chapter 1), governments around the world are coming to the view that substantial rises in the price of energy must be a central part of any policy for reducing carbon emissions. Therefore, energy prices are likely to rise even if there are no price rises due to oil peaking as governments impose energy or carbon taxes.

THE MECHANICS OF CLIMATE CHANGE

The science of climate change has been well documented and this section can only be a short summary. The Earth's surface is warmed by the sun and the Earth's temperature is controlled by the balance of energy gained from incoming short-wave radiation from the sun and the loss of energy which bounces back into space. Of the incoming solar radiation, roughly 30 per cent bounces back into space from clouds, atmospheric aerosols and bright, reflective parts of the Earth's surface. The remaining 70 per cent of the incoming radiation is absorbed, mostly by land areas and the oceans. This 70 per cent, however, cannot stay permanently absorbed, otherwise the Earth would have continually warmed up through its existence. It is re-emitted primarily as long-wave, infra-red radiation back into space. Some of this long-wave radiation is absorbed by water vapour and by 'greenhouse gases' that exist in the atmosphere. The principal greenhouse gas is carbon dioxide, but the principal absorbing agent is water vapour. The effect of this absorption of the re-radiated energy is to produce another round of re-radiation, this time back to the Earth's surface, where it is absorbed once again. This is the 'greenhouse effect' and is totally natural: it is what maintains the Earth's average temperature at +15°C rather than below freezing.

The increased concentration of greenhouse gases act as a 'blanket' that traps more of the outgoing long-wave radiation and returns it to Earth. It is this enhanced greenhouse effect that causes the concern.

The relatively short period since the industrial revolution has seen significant increases in the emissions of the 'greenhouse gases', especially carbon dioxide and methane, mostly from the combustion of fossil fuels, which adds to the concentrations already in the atmosphere. Evidence from Antarctic ice cores suggests that atmospheric concentrations of carbon dioxide (CO_2) were fairly constant over the 1000 years before the industrial revolution. One thousand years ago, concentrations of CO_2 (measured by ice core samples) were 280 ppm (parts per million) and the concentrations were the same around 1800 AD, whereas today they are some 385 ppm and growing at some 1.5 ppm every year. Similar patterns have been found for concentrations of other strong greenhouse gases such as nitrous oxide (N_2O_2) and methane (CH_4). Longer historical records from ice cores suggest that carbon dioxide levels are now at their highest levels in the past 400 000 years.

No one disputes the fact of temperature rise in the last 100 years or so. No one disputes that CO_2 is a greenhouse gas, and only a few scientists dispute that it has an enhanced 'greenhouse effect'. There is a strong majority view on climate change. Majorities are not always right but many major associations of scientists have adopted similar positions, including the Royal Society, the US National Academy of Sciences, the American Meteorological Society, the American Geophysical Union and the American Association for the Advancement of Science.

THE ANTICIPATED EFFECTS OF CLIMATE CHANGE

The IPCC's 2001 report suggests a range of mean temperature changes of 1.4°C to 5.8°C by 2100. It is important to remember that these projections are based on a range of different scenarios and the IPCC does not comment on which scenario is most likely. They also represent global mean temperatures and there are likely to be large geographical variations. The most certain effect is sea level rise due in the main to the thermal expansion of the oceans. The IPCC projects a mean global sea level rise of 20 to 88 centimetres by 2100. Again there will be local variations but any increase in sea level of this magnitude would threaten low-lying islands and coastal regions in many parts of the world.

The IPCC believe that the frequency of extreme weather events such as cyclones, hurricanes, tornadoes, ice storms, blizzards, rain storms and heat waves will increase and Sir John Houghton (co-chair of the Scientific Assessment Working Group of the Intergovernmental Panel on Climate Change and formerly chief executive of the Meteorological Office) has declared that 'these probably constitute the most important element in climate change impacts'.

Box 2.1 The Intergovernmental Panel on Climate Change

Recognizing the problem of potential global climate change, the World Meteorological Organization (WMO) and the United Nations Environment Programme (UNEP) established the Intergovernmental Panel on Climate Change (IPCC) in 1988. It is open to all members of the UN and WMO.

The role of the IPCC is to assess on a comprehensive, objective, open and transparent basis the scientific, technical and socio-economic information relevant to understanding the scientific basis of risk of human-induced climate change, its potential impacts, and options for adaptation and mitigation. The IPCC does not carry out research nor does it monitor climate-related data or other relevant parameters. It bases its assessment mainly on peer-reviewed and published scientific/technical literature.

Source: www.ipcc.ch (accessed 20 July 2006)

Data from insurance companies indicate a five-fold increase in the number of weather related events in the 1990s compared to the 1950s. Economic losses are estimated to have risen by a factor of 10 in the same period (at constant prices) (Allianz Group and WWF, June 2005).

There may also be impacts on human health. The IPCC suggests that:

- deaths associated with abnormally high summertime temperatures may rise;
- mortality and morbidity due to changes in the availability of drinking water are more likely with future warming;
- higher temperatures will decrease water availability in some parts of the world;
- vector-borne diseases will increase as regions warm.

Global warming could bring about ecosystem change and hence changes in the populations of species. The IPCC cites particular ecosystems at risk: glaciers, coral reefs and atolls, mangroves, tropical forests, polar and alpine ecosystems, prairie wetlands and remnant native grasslands. Other effects that could have economic impact include migration of species to new areas; for example, the North Sea off the UK has already experienced an increase in the number of species including squids, and their predators, dolphins, that would normally be expected to inhabit more southerly waters.

Impacts on agriculture are debated. Some regions could gain from a warmer climate and the fact that higher CO_2 levels enhance crop growth. Some regions, notably the poorer parts of the world, will lose because of changes in precipitation and higher temperatures. Another impact could be the large migrations of populations from seriously affected areas.

Many experts have emphasised the ability of global warming to generate 'large surprises', or what the IPCC refers to 'large scale singular events'. Climate change models generate results that suggest fairly smooth changes, which can be reassuring as they suggest that there is time to adjust and adapt to the new conditions. Change, however, could be non-linear and abrupt and there are many examples of non-linear behaviour in the Earth's climate system. The kinds of surprises that are considered possible include:

Reversal or shut down of the ocean thermohaline circulation (THC). The THC refers to deep ocean currents that move heat and water between the oceans. For example the Gulf Stream brings warm surface water from the Gulf of Mexico to the northeast Atlantic and returns cold deep water to the south Atlantic. The Gulf Stream maintains Europe's average temperature about 8°C higher than it would be otherwise. The fear is that additions of substantial amounts of freshwater from ice melts in the North Atlantic will reverse or switch off the vertical changes in the ocean's waters. European summers would heat up and winters could become very much colder. Some theorists say this change could happen within a decade.

Disintegration of the West Antarctic ice caps, which would alter the south Atlantic Ocean circulation and produce sea-level rise on a more dramatic scale than the increase due solely to thermal expansion, with an increase of several metres possible.

Melting of the permafrost. Methane gas mixed with water as ice exists in very large quantities in solid form below the permafrost. High levels of temperature change could release the methane, a powerful greenhouse gas, producing a 'runaway' acceleration of warming.

Acidification of the oceans, which would change ocean life dramatically. There is some evidence that this is already happening (Royal Society, 2005).

There could be significant regional effects that would have global consequences. These include major changes to the Amazon rainforest, desertification of large parts of Africa and major changes to the Indian monsoon. The scale of the effects of these threats, should they occur, are clearly very large. What is uncertain is their probability.

THE COSTS OF CLIMATE CHANGE

In considering the costs of global warming, the IPCC have considered possible objectives for stabilising carbon dioxide concentrations at various levels and the median one is 550 ppm. Estimates of the costs of stabilising carbon dioxide concentrations at this level by 2100 range from $2 trillion to $17 trillion in present value terms, that is the equivalent of spending this sum of money once and for all today. Expressed as an annual flow, the sums are $78 billion to $1141 billion per annum. These sums should be compared to the world's annual GNP of $35 trillion. Therefore annual expenditures would be 0.2 to 3.2 per cent of annual current income.

A House of Lords Select Committee report (House of Lords, 2005) suggests that the UK faces 'one off' costs equal to £60 to £400 billion, or an annual burden of £3 to £16 billion per year for the next 50 years.

The Stern Review (Stern, 2006) estimated that the cost of doing nothing to mitigate climate change could be equivalent to losing between 5 and 20 per cent of GDP each year. Investing about 1 per cent of GDP, however, would reduce greenhouse gas emissions and avoid the worst impacts of climate change.

ADAPTATION VERSUS MITIGATION

Most of the public debate and policies on climate change are about mitigation rather than adaptation, but the global economy will have to adapt to climate change, whatever magnitude it turns out to be. It is prudent to take action to mitigate the effects by reducing emissions, and this is the focus of government policy in most countries.

The global economy is made up of individuals, companies and governments as well as non-governmental organisations. All of these will have to make decisions about both mitigation and adaptation, decisions that will have an economic effect at both the micro and macro level.

The options open to companies to mitigate climate change include:

- improved energy efficiency;
- greater use of renewable or low-carbon energy sources;
- reduction in travel and transportation;
- implementation of carbon management programmes.

THE UK POLICY FRAMEWORK AROUND GLOBAL WARMING

UK government policy towards climate change has evolved since 1992. The original 'Rio' target (so-called as it was agreed at the UN Conference on Environment and Development – 'the Earth Summit' – in Rio de Janeiro in 1992) called for carbon dioxide emissions in 2000 to be no greater than in 1990. It was a voluntary agreement with no penalties for non-compliance. In the event, the UK was one of the very few countries that complied with the Rio targets, though not through policy design but rather through the choices made by newly privatised utilities, who made a large-scale switch from coal to natural gas.

The current UK 'Kyoto' target is determined by the EU burden-sharing agreement. In other words, it is the EU that has to comply by international law with its overall target. Non-compliance by individual EU member states with the burden-sharing agreement is a matter for internal EU law. Additional targets that have no legal compliance requirements are the UK government's 1997 manifesto commitment of 20 per cent reduction in carbon dioxide emissions (relative to 1990) by 2010, and its long-term aspiration of a 60 per cent reduction in carbon dioxide emissions.

SUMMARY

Global climate change is happening. There may be some debate about its scale, its long-term impact and the fundamental causes, but there is no question that it is happening. Although our understanding of climate change and the impact of human activity on the global climate is still limited, governments throughout the world recognise the problem and have initiated local, national and supra-national policies and programmes to mitigate climate change. Given that the emissions of carbon dioxide that predominantly come from the combustion of fuels is believed to be a major contributor to climate change, it is inevitable that government policies will attempt to limit their fossil fuel use and reduce carbon emissions through improved energy efficiency and the increased use of low carbon energy systems.

With pressure from government and the growing need to demonstrate responsible behaviour in all areas especially the environment, all organisations need to put in place strategies to address the climate change issue. Possible strategies range from a minimal approach driven through the need to comply with regulations through to a response which seeks to gain competitive advantage. Several leading companies, notably BP and GE, are following the latter route by focusing on climate change in advertising and developing products and services that contribute to mitigating greenhouse gas emissions.

CHAPTER 3

Legislation and Regulation

Increased pressure on governments around the world to respond to local and global environmental concerns have led to a significant increase in the amount of environmental legislation and regulation that affect all energy users. This chapter outlines the major current and foreseen regulations that impact on energy and utility use. It concentrates mainly on the UK regulations, all of which are now driven by EU legislation.

THE EVOLUTION OF UK POLICY AND REGULATION SINCE KYOTO

The UK policy on climate change has evolved through several steps since the signature of the Kyoto Protocol in 1997. The steps were as follows:

Kyoto Protocol, 1997

- UK committed to 12.5 per cent reduction in greenhouse gas emissions from 1990 baseline by 2008–2012.

Climate Change Levy

- Imposed significant average energy cost increases.

EU Emissions Trading Scheme, 2005

- Department of Trade and Industry (DTI) expects UK electricity prices to rise 3 to 6 per cent.

EU Energy Performance of Buildings Directive

- Sets new standards for the construction and refurbishment of buildings.

EU Directive on Energy End-use Efficiency and Energy Services

- Sets mandatory targets for public sector energy efficiency savings.

Each of these policies will be explained in turn below.

THE KYOTO PROTOCOL

The Kyoto Protocol, which was signed in 1997, and which is the most often talked about international law in connection with global warming, is actually an amendment to the United Nations Framework Convention on Climate Change (UNFCCC). The Protocol divides the world into Annex I and Annex II countries and requires Annex I countries, essentially the developed world, to commit to reduce emissions of carbon dioxide and five other greenhouse gases, or engage in emissions trading if they maintain or increase emissions of these gases. The protocol came into effect on 15 February 2005 after a long period in which it appeared as if insufficient countries would ratify it. The United States and Australia are the major countries which have refused to ratify the protocol.

The Kyoto Protocol can only be a first step in any international process to substantially ameliorate global warming, and in late 2005 agreement was reached in Montreal that the parties would negotiate an extension to the Kyoto Protocol.

Under the Kyoto Protocol the overall EU target, which has been allocated between EU countries, is 8 per cent, and the UK is legally committed to a 12.5 per cent reduction in greenhouse gas emissions from the 1990 baseline by 2008 to 2012.

CLIMATE CHANGE LEVY

Following the Kyoto Protocol, the UK government introduced the Climate Change Levy (CCL), an additional tax on businesses that is added to energy bills. The revenue raised by this tax is redistributed back to businesses by reducing the level of National Insurance contributions (employer-paid social security), so that the measure is described as 'revenue neutral'. However, the degree of neutrality depends on the ratio between energy expenditure and the number of employees. A business with many employees and a small energy spend benefits from this levy, whereas a business with a large energy bill and a small number of employees is penalised. Some of the revenue raised is also 'recycled' through the Carbon Trust, the independent agency introduced to promote the shift to a low-carbon economy.

Electricity generated by renewable energy is exempt from the levy, as is fuel used in combined heat and power (CHP) plants that meet certain quality criteria ('Good Quality CHP'), as certified by the CHP Quality Assurance scheme (CHPQA). Fuels used as feedstock, and electricity used in smelting processes such as aluminium smelting are also exempt. Petrol, diesel, and road fuel gases are also exempt from the CCL, as are 'waste' fuels from domestic, agricultural and commercial refuse, sewage and landfill gas, and bio-fuels. Fuels used for electricity generation or non-energy purposes are also exempt, but there may be additional form-filling to gain exemption approval for non-energy use of fuel. Renewable energy and energy from certain specified CHP plant is also exempt.

Forty-four business sectors (now reduced to 42) originally negotiated an 80 per cent reduction in CCL through their relevant sector trade associations by signing up to energy saving targets set up in agreements with the government. These targets are mainly in the form of relative consumption compared to production levels and are on a reducing scale with milestones every two years. The CCL was increased for the first time since its introduction in the budget announced in March 2006.

EU EMISSIONS TRADING SCHEME

The European Union Emissions Trading Scheme (EU ETS) commenced on 1 January 2005. The first phase runs from 2005 to 2007, with the second phase running from 2008 to 2012 to coincide with the first Kyoto commitment period. Further five-year periods are expected to follow.

The scheme works on a 'cap and trade' basis. EU Member State governments have set an emissions cap for all installations covered by the scheme. Each installation has been allocated allowances for the particular commitment period. The number of allowances allocated to each installation for any given period (the number of tradable allowances each installation will receive), is set down in the National Allocation Plan (NAP). Installations that reduce emissions below the allowance granted in the NAP are able to sell the surplus emissions on the EU-wide market. Installations which have emissions greater than their allowance have to buy emissions on the market or pay a fine of €40 per tonne in the first phase and €100 per tonne in the second phase.

During 2005, the price of carbon reached €30 per tonne for a short period but in 2006 it fell to less than €10 per tonne before returning to about €20 per tonne.

For a company that is an EU ETS registered site, the value of the carbon saving needs to be considered in any investment analysis for fuel-saving projects. This brings greater potential benefits, but also an added dimension of risk to any efficiency project

The EU ETS brings a considerable burden of record keeping and reporting, and some companies may want to outsource the administration. Failure to comply with reporting requirements will lead to a statutory notice from the Environment Agency and ultimately, if this is not responded to, fines and legal actions. Companies will have to improve the quality of monitoring and predicting of emissions in order to comply with the scheme requirements and assess the potential risks and rewards. For many companies, entering the world of trading has been a new experience with a steep learning curve, and a variety of routes to trading through intermediaries have emerged.

Given that the allocation of carbon dioxide is fixed in the National Allocation Plan, and the electricity generators have robust targets, it was perceived at the start of the scheme that the price pressure on EU ETS carbon dioxide was likely to be upwards, at least until phase 2 starts in 2007–2008. The sudden unexpected price falls in May 2006 shows that the new carbon market is extremely volatile and represents an added dimension of risk to any investment case. In the second stage the EU ETS will be opened up to allow carbon dioxide emissions from Clean Development Mechanisms/Joint Implementation (CDM/JI) projects (see Box 3.1) to be traded. This should lead to additional supplies of carbon dioxide becoming available from CDM/JI projects in developing countries and the former countries of the Soviet Union. Following the 2006 Energy Review it has been confirmed that the EU ETS will be extended to smaller energy consumers and other sectors including retailing.

Box 3.1 CDM/JI projects

The Clean Development Mechanism (CDM) and Joint Implementation (JI) are two of the so-called 'flexible mechanisms' of the Kyoto Protocol (the third being emissions trading). The mechanisms are project based because they are meant to promote projects in other countries that result in reduced emissions of greenhouse gases compared to what would have happened without the project. In the case of JI projects, both countries involved must have a reduction target under the Kyoto Protocol, whereas CDM projects occur in countries without an emission reduction target (that is, they must be a 'non-Annex I country'). The reductions in emissions have to be certified and the certified emissions can be bought and used by industrialised countries (Annex I countries) to fulfil their targets under the Kyoto Protocol. A share of the proceeds goes into a fund to pay for measures to alleviate climate change impacts in developing countries.

The second stage of the EU ETS will allow companies in the trading scheme to purchase certified emissions resulting from CDM/JI projects.

Source: Can Europe, www.climnet.org/EUenergy/CDM.htm (accessed 11 December 2006).

EU ENERGY PERFORMANCE OF BUILDINGS DIRECTIVE

Nearly 50 per cent of the final energy consumption in the UK and 40 per cent of the energy within the EU is accounted for by buildings. In 2003 The EU's Energy Performance of Buildings Directive became law. The principal objectives of the directive are:

- to promote the improvement of the energy performance of buildings within the EU through cost effective measures;
- to promote the convergence of building standards towards those of Member States that already have ambitious levels.

The practical effects of the directive, which came into force in the UK in January 2006, will include:

- the need to monitor energy use more which will require additional investment in metering to provide information at the necessary level;
- the need to issue energy performance certificates for buildings. Research suggests this will lead major property investors to sell low performing buildings and to invest in better performing buildings, which could have an effect on property prices (ACE, 2005).

Most of the requirements of this directive will, in the UK, be enforced through modifications to the building regulations.

Building regulations Part L

In April 2006 the UK building regulations were considerably tightened with respect to the energy performance of buildings in order to comply with the EU Energy Performance of Buildings Directive.

The aim of the regulations is to reduce the overall energy consumption as measured by the total annual carbon dioxide emissions (kg/m^2 of gross floor area). For new-build non-domestic buildings the target is to reduce the energy consumption by 20 per cent for air-conditioned and mechanically ventilated buildings compared to the previous regulations, and to make a 10 per cent renewable energy contribution where appropriate.

The regulations establish certain standards for building elements including maximum permitted air tightness, minimum seasonal efficiency of boilers, minimum lighting efficiencies and maximum specific fan power for mechanical ventilation systems. All design factors have to be fed into a model and compared to a notional building of the same size and shape conforming to the 2002 building regulations. For the majority of non-domestic buildings the energy reduction needs to be 28 per cent compared to the same notional building.

Meeting these targets, especially the 10 per cent renewable contribution, will be challenging for many buildings unless an explicit low-energy, low-environmental impact approach is taken at the very early stages of the brief. Trying to 'add in' energy efficiency at a later stage in the design will be difficult. Technologies such as solar thermal, micro-wind power, building integrated photovoltaics (BIPV) and ground source heat pumps (GSHP) should receive a boost from the new regulations. Unlike previous regulations, the new Part L requires that the building be tested prior to handover to demonstrate that it achieves the required carbon emission target.

Another revision of the building regulations is planned for 2010 and it is likely that additional measures will be included to further improve the energy performance of new buildings.

Although most organisations will not be involved in specifying and constructing new buildings, it should also be noted that the new regulations also apply to refurbishments of buildings with an area greater than $1000m^2$. Furthermore, reviews of energy performance are required every five years. Buildings greater than $1000m^2$ occupied by public organisations or visited by large numbers of the public will have to display energy performance certificates showing an energy rating for the building. Although they have not yet come into force there will almost certainly be a system of regular mandatory inspections for air conditioning systems and boilers.

As well as the regulations directly associated with climate change, two other sets of regulations affect energy use. These are the Integrated Pollution Prevention and Control (IPPC) regulations and the Operating and Financial Review (OFR) regulations. These are now discussed in turn.

INTEGRATED POLLUTION PREVENTION AND CONTROL (IPPC)

The EU has a set of common rules on environmental permitting for industrial applications, which are set out in the IPPC Directive of 1996. In essence, the IPPC Directive is about minimising pollution from various point sources throughout the EU. All installations covered by the directive are required to obtain a permit from the appropriate authority in their EU country; without a permit the installation cannot be operated. The permits are based on the concept of best available technique (BAT). In many cases BAT means quite a radical environmental improvement and a significant investment in new plant and equipment and so to give firms time to adapt to the new requirements the EU allowed an 11-year transition period. In the UK the adoption of IPPC in different sectors is following a schedule set down by the Environment Agency (EA), the permit-issuing authority, and the EA has published guides, which detail BAT for several industrial sectors.

EU DIRECTIVE ON ENERGY END-USE EFFICIENCY AND ENERGY SERVICES

In December 2005 the EU adopted the Directive on the Promotion of End-use Efficiency and Energy Services to promote the implementation of cost-effective energy efficiency measures and improve the efficiency of energy use. Coming into effect over nine years starting in January 2008, the directive will provide targets, mechanisms, incentives and institutional, financial and legal frameworks to remove existing market barriers and imperfections that reduce the uptake of energy efficiency measures. The proposal sets out clear, mandatory, targets for annual energy savings at the Member States level and for the share of energy efficient procurement for the period 2006 to 2012. Strong incentives will be given to ensure that suppliers of energy offer a certain level of energy services.

Perhaps the item that will have the most effect is Article 13 of the directive, which states that consumers should be provided with meters that accurately reflect the customer's actual energy consumption and the time of use. It will require energy bills to be presented in a 'clear and understandable' way and issued frequently enough to enable customers to regulate their own energy consumption. This will have implications for metering technology and utility companies' billing practices, in both the domestic and the business sectors. Bills will also have to show current prices and actual consumption, a comparison with previous years' consumption, and an indication of a benchmarked use for an energy user in a similar category in clear and understandable terms. As with other directives there will be nationally agreed transition periods but this directive, when it comes into effect in the UK, will further increase pressure on energy suppliers and energy consumers to monitor and manage energy more effectively.

OPERATING AND FINANCIAL REVIEW (OFR)

The Operating and Financial Review (OFR) regulations came into force in the UK on 22 March 2005. They took effect for company financial years beginning on or after 1 April 2005 and apply to all listed companies (including UK companies listed on the New York Stock Exchange and NASDAQ).

The OFR requires the directors of listed companies to produce in the annual report a 'description of the principal risks and uncertainties facing the company'. The OFR sets out the directors' strategies for achieving success and looks at performance of those strategies. As well as past performance, the OFR needs to look at 'the main trends and factors which are likely to affect the company's future development, position and performance', that is, it should be forward-looking. The OFR regulations are quite specific in requiring companies to consider three non-financial issues: environment, employees and community and social issues. Performance has to be reported in the form of quantifiable key performance indicators (KPIs), and recommended KPIs in the environmental area include:

- water use
- energy use
- waste
- global warming contribution
- ozone-depleting substances
- environmental spillages.

Clearly environmental performance and risks will be a large part of the OFR, including the potential impact of global climate change.

The OFR has to be audited by external auditors and failure to complete the OFR according to the regulations can be a criminal offence by the directors. The OFR is not expected to replace corporate social responsibility (CSR) reports, but clearly there will be overlap between the OFR and any CSR report.

Although little noticed, the OFR has the potential to make a significant impact on company's reporting of environmental issues.

In the United States the Sarbanes-Oxley Act requires companies to report operational risks that may materially affect the financial performance of the company, which may include energy prices and energy-related risks.

SUMMARY

Over the last decade energy users have seen a marked increase in the number and the complexity of regulations that impact on energy consuming operations. Given the imperative of global climate change it is likely that this trend will continue, but at the same time many organisations have downsized specialised energy support staff. Keeping up with, understanding and complying with the reporting regulations can be a full-time job and the outsourcing of administration of the regulations is likely to be a growth area, although of course it is not possible to outsource the ultimate responsibility for compliance which sits with company directors.

The regulations described above are only those that relate directly to energy or climate change. Any organisation operating any type of energy using plant will also be subject to legislation and regulations in areas such as health and safety and the environment.

Summary of Part 1

This section set out to explain why energy is an important issue and why all of us need to be concerned about it. It has summarised the energy-related issues facing all of us at a global level: oil peaking and climate change. The exact date of peak oil production may only be known after the event, but there is no dispute that it is inevitable, and may occur within ten years. The global geopolitical and economic effects that will result from the world having to reduce its addiction to oil, and eventually give up altogether, cannot be underestimated. They will affect all of us as individuals as well as our companies and organisations.

Climate change is also now an accepted fact. Even if there remains doubt about the causes of global warming being solely attributable to man-made emissions of greenhouse gases, national policies and programmes in many countries are now firmly based on that linkage. Those policies and programmes are unlikely to be reversed in the foreseeable future. Even if fossil fuels were in greater abundance, delivering energy to the entire population at the standards we in the developed world enjoy today by fossil fuels would lead to a massive increase in carbon dioxide emissions and hence accelerated global warming.

PART 2

The Energy Management Process

Second law of thermodynamics

1. *heat flows spontaneously from a hot body to a cool one;*
2. *one cannot convert heat completely into useful work;*
3. *every isolated system becomes disorderly in time.*

INTRODUCTION

As we have seen in Part 1, energy is a crucial issue and one that is likely to increase in importance over the coming years as the effects of oil peaking and the pressure to reduce greenhouse gas emissions intensify. It is worthy of management time in all organisations. Clearly the optimum amount of time and resource applied to it will be a function of many factors, including the size of the organisation, its relative energy intensity, interest from senior management and market perception. It is likely, given that we have just emerged from a long period of relatively low energy prices and prices have increased dramatically, that whatever level of time and resources have been, or are currently being applied to the energy problem, they are insufficient to face the current and forthcoming challenges.

The basics of energy management are the same for all organisations, whatever their energy intensity or energy spend, and it is important to stress that energy is a management problem and not a technical one. It needs to be addressed in the same way as other management issues and not driven by technological issues or technologists. It has technical aspects, of course, but it should always be a management issue and treated as such.

An organisation's energy cost is a combination of the price paid for energy and the volume of energy used. Energy price, or energy supply management, is traditionally the concern of energy procurement, whilst the volume consumed is the concern of energy demand management. Often 'energy management' is used to mean simply energy demand management, but true energy management looks at both the supply and the demand sides of the equation. Here we define energy supply management as 'the management of energy price and price risk' and energy demand management as 'the management of energy consumption or volume'. One of the barriers to improved energy management has been organisational structure, whereby energy supply management is controlled by the procurement department, and energy demand management by a technical department. By separating these two functions many organisations restrict themselves; combining the two sides of demand and supply into one organisation is likely to bring benefits.

The purpose of Part 2 is to set out the basic processes and activities required in any energy management programme, covering both energy supply and energy demand. For organisations considering outsourcing of energy management it is essential to understand the basic processes so that they can effectively specify the scope of any outsourcing exercise.

On the energy demand side the technical principles are universal, but prevailing energy prices, tariff structures and the cost and availability of technology and capital will vary from country to country and drive what is economically and practically possible. On the other side of the equation, energy supply, there are more differences between countries and markets as the energy markets, particularly power and gas markets, are driven by regulation with different markets operating under different regulatory regimes. Across the world over the last 20 years there has been a significant deregulation (or more accurately 're-regulation'), of power and gas markets as governments, led by the UK in the late 1980s and 1990s, swept away old monopolies and introduced market-driven competition.

CHAPTER 4

Energy Management

Organisations and individual consumers are not concerned directly with the flows of energy through them, only with the useful outputs, and the total costs of producing those useful outputs. Figure 4.1 summarises the flow of energy through the organisation.

It is useful to go back to basics to fully understand the energy systems in a factory or building.

Energy *commodities*, gas, oil, coal and electricity, are purchased and bought into the organisation where they are transformed into *utilities* such as steam, compressed air, coolth (refrigeration) and usable power (at different voltages and quality levels) in *utility conversion systems* such as boilers, air compressors, chillers and transformers. Each item of plant has an operational *conversion efficiency* (which may vary over time as plant ages or as maintenance practices vary), which is defined as utility out/energy in. The utilities are then distributed to the point of end use through *distribution systems* consisting of pipes and cables. The distribution systems will also have an inherent *efficiency* (usually fairly high in well-maintained systems). At the point of end use the utilities are transformed into *useful work* such as the manufacture of components or the production of comfort conditions in a building (for example light, fresh air and warmth). At the point of use the utilities are converted to useful work with an inherent *efficiency*, depending on the nature of the process or space conditioning plant, and are used with a certain *effectiveness*. Effectiveness covers matters such as leaving heating and lighting on when the building is empty or leaving motors running when plant is idle – issues which are often labelled 'good housekeeping'. The level of effectiveness is usually controllable in some form, whether by automatic controls or management systems. Following the conversion of a utility to useful work there will be some *by-product streams* which, like the other efficiencies referred to above, are a consequence of the second law of thermodynamics. In the case of steam the by-product will be hot water (condensate), most of which can be recovered cost effectively. In the case of an electrical process the by-product will be heat (and noise). An inevitable by-product stream of any combustion process will be carbon dioxide, which contributes to global warming, as well as other emissions such as sulphur dioxide and particulates. When combustion equipment such as boilers or

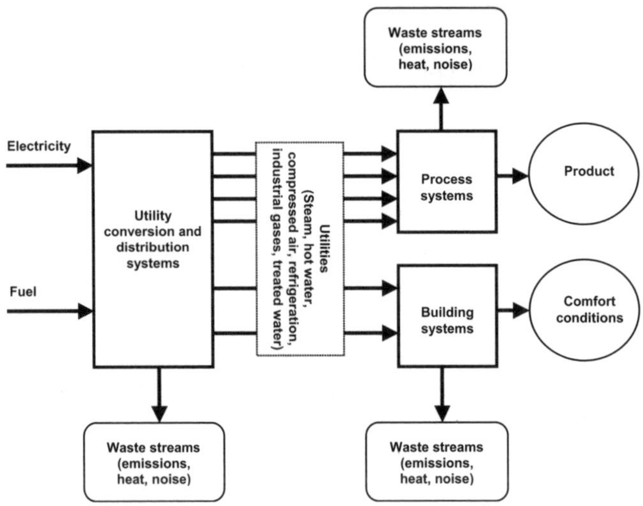

Figure 4.1 Schematic energy flows through an organisation

process heaters are used onsite, there is a direct emission of carbon dioxide. With the use of electricity there is an indirect emission of carbon dioxide that results from the process of generating electricity at the power station unless the electricity was generated from a non-fossil fuel source such as wind or nuclear power.

The other inputs to the total costs of the process, which is essentially one of energy degradation are:

- *capital* used to buy the plant and distribution systems;
- *operations and maintenance (O&M) costs*, such as bought in services, spare parts and consumables such as boiler chemicals;
- *O&M labour costs;*
- *management time* taken to procure the energy commodities;
- *management time* to manage the O&M processes
- *management time* to produce and utilise the information needed to run the system;
- *the costs of any by-products* that carry a disposal cost for example effluent charges, chemical disposal costs and (when the EU ETS or other emissions regulations apply) a cost associated with emission of carbon dioxide;
- *the costs of any permits or licences.*

The objective of total energy management should be to minimise the total long-term cost of utilities. This involves:

- minimising the unit price of bought in commodities;
- maximising/optimising conversion efficiencies;
- maximising effectiveness of end use;
- optimising capital requirements.

Total delivered energy cost (or utility cost) is made up of:

units of energy used × price per unit +

capital employed in energy plant and system to deliver energy +

O&M labour cost of staff to operate and maintain utility systems +

O&M costs to maintain utility systems +

disposal costs +

management costs +

permit or licence costs.

Failure to recognise the need to look at total cost of delivered energy can lead to management focusing on the wrong things, and in many organisations energy management has been, and still is, excessively concerned with the energy supply question: 'how can we buy energy at the cheapest price?' Energy managers focusing on the demand side ask: 'how can we maximise efficiency and effectiveness?' There is often a failure of accounting systems to recognise the true total cost as costs are split into different departments (the use of activity based costing – a set of accounting methods used to identify and describe costs and required resources for activities within processes – may go some way to correct this). The correct questions that should be being asked by senior management, in this order are 'how do we minimise end use energy requirements?' and 'how do we meet the end use requirements at the minimum total costs with the minimal environmental impact?'

Framing the questions in this way leads to a different approach to 'energy management' than that which is traditionally meant by the phrase. The questions that energy service companies and utilities should be asking in relation to their customers should be the same – but traditionally have been different.

ENERGY SUPPLY-SIDE MANAGEMENT

Cash flow can often be disrupted by movements in external factors such as exchange rates, commodity prices and interest rates potentially compromising a company's ability to invest. A risk management program, therefore, should have a single overarching goal: to ensure that a company has the cash available to make value-enhancing investments.

K.A. Froot, David S. Scharfstein and J. Stein (1994) 'A Framework for Risk Management', *Harvard Business Review* (November–December)

Energy supply management in the UK (particularly for electricity) has gone through three stages since the privatisation of the electricity supply industry. Prior to privatisation customers of all types simply had to take the prices and tariff structures that were on offer from the local supply company. The only way to save money was to ensure that the tariff charged was the most appropriate, and hence the cheapest, for the load pattern or 'shape' of the building or facility. The supply companies did not, as a general rule, spend any effort on assisting consumers to ensure they were on the best tariff and so many organisations with large portfolios of properties either employed staff or external consultants (tariff consultants) to check load patterns, change buildings to advantageous tariffs and validate bills.

As the impact of privatisation took effect (starting with 1 MW users, then 100 kW users and then all users in the UK) the emphasis in procurement switched to obtaining the lowest possible price per MWh through competition, usually in the form of tenders. These required users to assemble their consumption data and issue tenders to the competing electricity suppliers for their entire power supply. In the UK this became institutionalised into two annual rounds, April and October. Customers with limited internal resources began to use intermediaries to assemble data and run the tender process. The pressure was to achieve the lowest price per MWh (or therm of gas) and, due to the competitive nature of the market, the real price of electricity fell year on year. In almost all cases the customers contracted for their entire annual energy usage on one day. As a large proportion of the total demand was priced and contracted in a short period twice a year and power supply companies knew what to expect.

Since 2005 there has been a rapid and dramatic change in the way that the large power consumers buy power. Instead of nominating one day in advance when they will buy all their power for the year, customers are increasingly looking for flexible, long-term relationships with a supplier that gives them access to the same kind of trading and risk management tools used by the power companies themselves. These include buying a proportion of base load power on a long-term deal and being able to enter the market at will to take advantage of the best price and day-to-day swings in price.

Leading industry figures reported that the large user market had switched to this new procurement approach almost overnight. In April 2004 very few customers wanted this kind of service, whereas by April 2005, fully half of the large industrial and commercial market had switched to a new types of contracts including risk management products. Typically the customer is given direct access to the wholesale market, that is the trading desk, and the retail energy supplier charges a transparent margin for this service. This change in the energy purchase options available to organisations, along with the changed approach to risk management, have a direct impact on organisational structure and processes. It is no longer appropriate to contain energy procurement with a purchasing department. Energy purchasing has to become more of a strategic function that requires new structures and new controls to manage risks and costs. In response to the changed market, energy procurement teams also need new skills and a deeper understanding of electricity markets and power and gas trading.

As well as major changes in the way electricity is purchased, there are an increasing number of alternative supply options such as purchasing renewable, 'green' power from the supplier, as well as investing in on-site, embedded generating systems such as combined heat and power (CHP) or renewables. Renewables, with the exception of embedded wind power and solar thermal, are not generally economic; they do offer the potential for price certainty over a long period, and in an increasingly volatile market this, along with its CSR benefits, may be attractive.

BILLING ISSUES

A major activity for energy managers and consultants is the checking and validation of energy invoices. This has been vital over the last decade as utility billing in the UK has generally been atrocious. The low standard of billing would never be acceptable in any other business and probably stems from the legacy of the suppliers, often formed through mergers and acquisitions, and the difficulties of merging numerous different billing systems as well as their history in coming from a nationalised industry, 'take it or leave it' culture. This issue is being addressed by suppliers through investment in improved IT systems and ultimately could be eliminated through the use of intelligent, advanced monitoring and targeting systems that can provide real-time data and self-billing. It should actually be in the interest of the power supplier to have the customer install this kind of system for several reasons, including reducing billing errors, improving cash flow and customer satisfaction, producing more reliable consumption data, which can then be used to predict demand more accurately, thus reducing risks.

ENERGY SUPPLY RISKS TO BE MANAGED

Effective energy supply management requires detailed knowledge of your site or portfolio energy consumption, coupled with an agreed risk-management strategy. For large sites, and increasingly for smaller sites, consumption data will be in half-hourly intervals and the collection and management of this data to ensure its accuracy is essential for obtaining the best price. The risk management strategy needs to set the parameters for how much risk the organisation is willing to take in this area and establish appropriate guidelines for entering the markets as well as delegations of authority. Procuring all of your power or gas demand on one day has the effect of fixing the cost but may lead to missing opportunities to take advantage of swings in the price.

Other risk factors to be considered include the stability of the supplier, the effect of weather variations on demand, and the impact of regulatory changes and taxes such as the Climate Change Levy. In addition, energy supply management requires an understanding of the terms and conditions used in the power and gas industries.

To date, many organisations have not had in place the sophisticated energy information systems that are needed for proper risk management and in many cases consultants have been used to assist in the risk management process.

ENERGY DEMAND-SIDE MANAGEMENT

The corporate and national focus on controlling energy demand has waxed and waned in line with variations in the relative price and/or scarcity of energy supplies. It wasn't, however, until the 'oil crises' of 1973 and 1979 and the increase in energy prices that they brought about that energy management began to be standardised and codified into a recognisable and teachable discipline.

During the 1980s many organisations built capability and capacity in the energy management arena and many significant programmes were implemented across the economy with major results. During the 1990s, when competition in the energy supply markets resulted in lower energy prices, much of this capacity was eroded as companies downsized generally and in particular energy management specialists moved into energy procurement roles or other areas of activity altogether. Many organisations that previously had centralised energy management teams now have no specific organisation and energy is the part-time responsibility of an executive who has many other pressures.

With increasing energy prices and increasing environmental pressures, there is now a need to rebuild energy management capacity within end-user companies, within energy service companies, and increasingly within energy suppliers. In addition energy management has grown beyond demand-side measures to include distributed, or embedded, supply side technologies such as CHP, renewables as well as biomass projects. There has also been an expansion into carbon management, which takes a holistic view of an organisation's carbon emissions including those from transportation.

This chapter sets out the basic activities that are essential to effective energy management and explains the most commonly used enabling tools. Over time, despite changes in available technologies, particularly in the metering and information arena, there has been a tendency for energy management to become fossilised. It is generally recognised in all economies there is a large gap between the identified economic potential for energy savings and the implemented potential (Sorrell, 2005). To help address that gap, energy demand management requires refocusing on the objective, that is achievement of the economic potential, and that this best done by a reframing and re-focusing of the energy management process. Often energy demand management does not deliver its full potential because one or more of the essential elements of effective energy demand management are not present or are diluted. This section presents a model of energy management that works to deliver maximum benefits and has been well proven in many different types of organisation.

For every organisation, large or small, private or public sector, single site or large portfolio, the basic processes and enabling tools for managing energy use are the same. What varies is the way in which the tools are applied, the appropriate level of resources and the degree of in house (or outsourced) expertise required.

Identifying where an organisation is in the process is essential for managers, consultants, energy service companies and other agents of change. Many organisations currently need to re-evaluate, reinvigorate, replace or restructure existing energy management programmes that have withered or lost focus and momentum during the period of relatively low energy prices. Making this change may encounter resistance from remaining energy management staff but will help them to become more effective.

The basic activities are discussed in the text and split into three levels:

- senior management;
- operational management;
- specialist energy management.

Each of the commonly used tools are described in a relevant box at the end of the chapter.

THE ACTIVITIES FOR SENIOR MANAGEMENT

The activities for senior managers are:

1. recognise the opportunity;
2. recognise the risks (value at stake);
3. decide on the strategic response:
    a. do nothing;
    b. implement an energy management programme focused on cost saving;
    c. use energy as a strategic opportunity;
4. set policy;
5. establish an energy management programme;
6. establish targets;
7. ensure reporting procedures are put into place;
8. allocate resources;
9. design and deliver appropriate organisational changes;
10. link to other appropriate areas, for example, corporate social responsibility or purchasing.

Each of these are now discussed in turn.

Recognise the opportunity

Why should a company implement an energy management programme? As well as non-financial, corporate social responsibility (CSR) related benefits, and offsetting the current and future price risks, there is significant potential to reduce energy costs.

Before attempting to quantify this potential we need to briefly look at what we mean by 'potential', as differing definitions are used, sometimes without clearly stating which definition is being applied.

Firstly there is a technical potential. This is the potential reduction in energy use or cost that could be achieved if all technically possible measures were implemented. As a purely technical potential, with no regard for economics, it is of little interest to most businesses or even public sector organisations that have to operate in an economic context.

Secondly there is economic potential. This is the potential reduction in energy use or cost that could be achieved if all technically possible and economically viable measures were implemented. The important point, often forgotten by academics and other outside analysts when considering the issue of potential, is that what is economically viable is defined by the investing firm. Thus for one firm, opportunities with a three-year payback may be acceptable, whereas for other firms opportunities have to achieve a one-year payback to be considered viable. Only the investor can determine what is economically viable, and indeed one attraction of outsourcing for individual firms as well as the economy as a whole, is that energy service companies (ESCOs) may have longer payback period criteria than their client companies, thus increasing the economic potential to make savings. This can be particularly true for utility-owned ESCOs, who may be able to compare investments in energy saving projects with paybacks of three to four years with energy supply projects that typically have much longer paybacks, meaning it may be more profitable to invest in demand-side management measures than new generation or transmission and distribution systems. Indeed, investing in such projects may actually remove or reduce the need to invest in new infrastructure projects.

Many estimates of the economic potential for energy saving have been made. Clearly the size of the potential is determined partly by the price of energy, an increase in energy prices automatically increases the size of the potential. Similarly reductions in the price of appropriate energy-saving technologies, or an improvement in their capabilities, increases the economic potential.

The potential for energy saving can be split into the following categories:

- no cost
- low cost
- high cost.

No-cost measures are those that involve changing behaviour, for example switching PCs and lights off at night, shutting open doors and windows, etc. Low-cost measures involve items such as adjusting existing controls (which could effectively be no-cost for many organisations) and draught proofing. High-cost measures require investment but still have an attractive payback. A rule of thumb is that there is 5 to 10 per cent savings potential through no-cost measures, 5 to 10 per cent through low-cost measures, and 5 to 10 per cent through high-cost measures. The cut-off points between low cost and high cost will be industry and company specific. Given the recent increases in energy price and the reduction in energy management capability, it is likely that these kinds of potentials exist in many organisations.

When considering the potential for energy demand-side savings it is important to think beyond the proportion of total costs represented by energy costs. Although the total energy cost may be a small proportion of total costs, suggesting that it is not worthy of much management time and attention, the best comparison to make is to compare the effort needed to make energy savings with the effort needed to make the same profit through increased sales. As profit margins in many businesses continue to be squeezed this should be fully recognised.

Some analysts have identified that taking a different approach to conceptualising the energy problem, and a different approach to engineering utility-related equipment, can actually produce savings far greater than 20 to 30 per cent. The results gained with this new approach ('information-based engineering' – see Chapter 8) are exciting and far from being applied in most organisations. This kind of radical approach is also well suited to ESCOs because the savings are greater, allowing quicker capital recovery and a more attractive proposition to put to the customer.

Recognise the risks

Senior management need to consider the risks associated with their energy use, including:

- risk of price increases;

- risk of brand damage;
- risk of exceeding emissions allowance (and subsequent costs);
- risk of losing production (energy efficiency can improve throughput and quality);
- risk of becoming uncompetitive if energy prices rise significantly as a consequence of the problems oil peaking and climate change.

Energy risks will be considered further in Chapter 6.

Decide on the strategic response

There are three possible strategic responses to the energy issues:

- do nothing;
- implement an energy management programme focused on cost saving;
- use energy as a strategic opportunity.

These are all valid responses as long as the decision is taken explicitly whereas in many organisations the decision is often taken implicitly. Part I argued that with the combined pressures of oil peaking and climate change, the 'do nothing' option is no longer an option, at least for most organisations.

The second option, implementing an energy management programme focused on cost saving, is likely to be the most frequent choice. If implemented correctly it will be profitable and produce cost-effective savings as well as reductions in emissions and other less tangible benefits. The activities and tools described below apply to this response. Viewing energy as a strategic opportunity means moving beyond simple cost saving and brings into account factors such as risk reduction, customer perception and brand image. It is based on the premise that being seen to reduce energy consumption and invest in low carbon energy schemes (often high profile although small in output) will have beneficial, indirect effects on the business beyond simple cost-cutting. The indirect effects can include:

- improved customer perception;
- increased sales;
- increased attractiveness to young people entering the job market;
- increased employee satisfaction.

Set policy

It is important for senior management to establish an energy policy. The ideal policy will be short and should include specific targets that can be updated as progress is made. A sample energy policy is shown in Box 4.1.

Box 4.1 Sample energy policy for an industrial organisation

Energy policy

The company is committed to purchasing and using energy in most efficient, cost-effective and environmentally responsible manner possible. Towards this end, the company will:

Improve energy efficiency by implementing effective energy management programmes that support all operations and customer satisfaction while maintaining a safe and comfortable working environment.

Objectives

- Benchmark all premises by December 2006.
- Reduce energy use per output by 15 per cent relative to the 2003 base year by 2008.
- Implement a programme of staff awareness and education on energy.
- Purchase 10 per cent of our energy use from renewable sources by 2008.

This policy applies to all of our premises.

Approved: CEO

Date: 20 July 2005

Establish an energy management programme

Senior staff should ensure that an effective energy management programme is established. For organisations beginning an energy management programme afresh the activities in the programme will be:

- establish monitoring and targeting (M&T);
- identify and implement quick win opportunities;
- identify investment opportunities;
- implement viable investment opportunities.

Responsibilities need to be assigned to members of staff and progress against milestones monitored.

For organisations with existing programmes the task for senior management is to identify and diagnose problems and restructure and refocus programmes to address those problems. External management-focused consultants may be helpful in this respect. In many cases existing energy management information systems will be inadequate and ineffective and need redesigning.

Establish targets

An overall target for energy efficiency should be set. As with all targets it should be SMART (specific, measurable, achievable, realistic, time related). An example of a SMART target would be, 'to reduce energy consumption per metre square of floor area by 10 per cent within three years'. Targets can be developed either from the top down or from the bottom up. Asking departments and subsidiaries for targets in a bottom-up exercise is likely to engage end-users more but has the danger that they will be understated. A top-down target has the danger of being unrealistic, which will only serve to de-motivate end users and energy management specialists. As well as a target for improved energy savings or improved efficiency, organisations opting for a more aggressive, strategic response may set targets for the use of carbon emissions (see Box 4.14 on carbon management on page 63), or the use of renewable energy (even though it is unlikely to meet normal economic criteria). Examples of targets in these areas include those set by the mayor of London, Ken Livingstone, in the Energy Strategy for London (GLA, 2004):

> London's renewable energy targets aim to generate at least:
>
> - 665 GWh of electricity and
> - 280 GWh of heat from 40,000 renewable energy schemes by 2010.

The targets are specific, measurable and time related. Whether they are achievable or realistic is debatable.

Ensure reporting procedures are put in place

Once energy monitoring and targeting (M&T) has been established reporting procedures need to be put in place. Periodic performance information needs to be sent to appropriate people, typically unit or cost centre managers who can control energy use, and summary data needs to be reported to top management. For cost centre managers the ideal frequency of this information will vary between situations but the general rule should be that performance should be reported as quickly as possible in order to allow people to see the effect of their actions and correct variances quickly. In a retail environment, reporting energy performance monthly may be sufficient but in a manufacturing environment daily or even in real-time may be more appropriate.

Allocate resources

An effective energy management programme requires resources, both human and financial. The allocation of resources will vary from organisation to organisation but, as a rule of thumb, experience from the 1980s shows that any organisation that has not invested in energy management can usefully and profitably invest at least 10 per cent of its energy expenditure in energy efficiency projects for three to five years consistently. This refers to investment in retrofit energy saving projects and excludes investment in major new energy infrastructure for replacement purposes.

Design and deliver appropriate organisational changes

Energy management requires an appropriate organisational structure. In the 1980s many organisations made energy managers responsible for achieving energy efficiency targets, but this is far from optimal. As energy use is directly affected by end-users the most appropriate, and most effective, organisational structure is to make site managers, departmental managers or cost centre managers responsible for energy use and achieving energy efficiency targets. As non-specialists they will require specialist technical assistance from an energy manager or outside consultants but the energy manager can only be a service function. Cost centre managers will also require appropriate energy performance reporting systems which are easy to understand and use. Although at the 'macro' level energy is somehow 'special' because of its essential nature, at the 'micro' level of the organisation it is important to manage energy like any other resource.

Link to other appropriate areas, for example, corporate social responsibility or purchasing

The energy management organisation and processes must be linked to other functions within the organisation, particularly maintenance, purchasing and corporate social responsibility (CSR); all too often these links are not made or are ineffective due to organisational barriers. Energy use, particularly in buildings, is affected by the quality of maintenance of heating, air conditioning and lighting systems. If the maintenance function is not motivated to reduce energy, consumption results will be sub-optimal. In the longer term energy use is affected by purchasing decisions – many organisations are still purchasing equipment on a purely lowest first-cost basis rather than on a whole life-cost basis. As well as locking in higher than necessary consumption over the life of the equipment this can have a negative effect on employees' views of any energy programme. Part of the energy policy should address ensuring purchased goods and equipment are energy efficient.

Corporate social responsibility includes environmental responsibility and energy use causes a major, if not the major, environmental impact. Energy reporting must be properly linked to CSR reporting.

The tasks for senior management are to establish and maintain two linked but separate sustainable processes. Firstly, a process for achieving and then maintaining the minimum energy input for a given output from a process, sub-process or building. This is essentially the process of managing the existing plant and equipment to attain maximum effectiveness, that is, only using energy when it is really required; and maximum efficiency, what is commonly called 'good housekeeping' (although it goes beyond that). The second, parallel process is that of identifying, developing and implementing cost-effective energy cost saving or carbon reduction projects (see Figure 4.2). This entails modifying or replacing existing plant and equipment to improve the effectiveness and efficiency of energy use.

THE TASKS FOR OPERATIONAL MANAGEMENT

Operational management covers cost centre managers and site managers as well as energy management specialists. The tasks for the two levels, within the good housekeeping cycle and the project cycle, are described below.

Good housekeeping cycle

Identifying and maintaining minimum energy use

Identification of minimum energy use for a given process requires the use of a monitoring and targeting (M&T) system. A base year should be established and

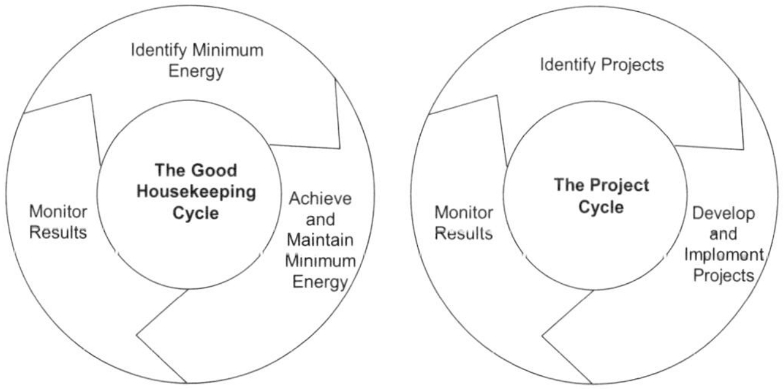

Figure 4.2 The two energy management processes

it is necessary to monitor performance regularly and feed back results to those people who can affect energy use. Ideally this would be done for each major energy cost centre. Usually when M&T is established variations in performance for a given level of output, production, or average temperature in the case of heating systems, are observed. Reducing this statistical variation will lead to reduced energy use (and often improved process control).

Having identified and achieved the minimum energy use it is necessary to maintain this performance. This is done by monitoring results and maximising staff awareness at all levels through the use of information, training and incentives.

Monitoring results

To engage and motivate staff the M&T results must be fed back to them in a timely manner, put into context by explaining what is at stake for the company and for the wider environment, and used by them to challenge their behaviour. Appropriate questions include: 'why did we use 20 per cent more energy to produce 500 tonnes of output this week than we did last week?' and 'what did we do this week that was different?' Encouraging a culture in which these questions are asked regularly, and the answers driven out and acted upon, is essential for successful management of energy within the good housekeeping level.

Motivation and awareness campaigns

Significant energy savings can be achieved by motivation and awareness campaigns designed to alter people's behaviour. Some organisations have reported savings of 20 per cent through awareness campaigns; the author has achieved savings of 6 per cent in a local authority portfolio. A large part of energy use is under the control of people, whether they are aware of it or not. Uniformity of behaviour helps to reduce the variations in energy use for any given activity, for example, for a given production output or number of hours a building is in use.

To achieve maximum success in an awareness campaign it is important to ensure that the all of the essential elements are incorporated, namely; training pitched at the right level to improve know-how of what to do and motivation, feedback on performance to improve motivation, and some incentive (which can be financial or non-financial). Figure 4.3 shows these elements.

Without all these elements being addressed in the design of an awareness programme the results will be sub-optimal. Often awareness campaigns consist purely of generic advice that is not helpful, with no feedback on performance.

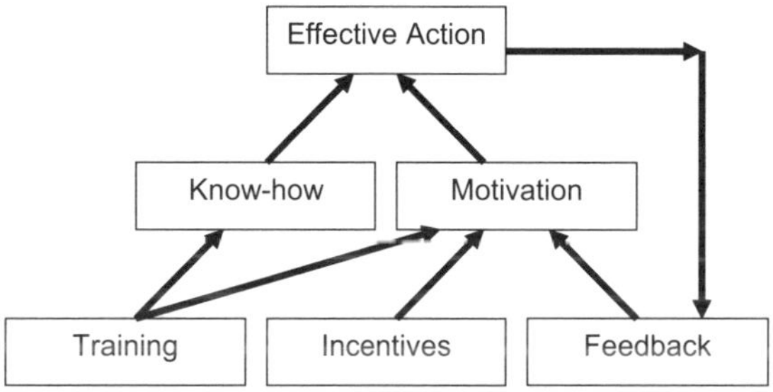

Figure 4.3 Elements of a successful awareness programme

The importance of implementing good housekeeping before investment programmes

There are sound reasons for implementing good housekeeping and achieving the minimum energy use prior to beginning an investment programme. Experience shows that there will nearly always be good potential for making savings by improving good housekeeping, somewhere between 5 and 10 per cent as a general rule, although individual examples in excess of 20 per cent have been achieved. Achieving the maximum potential savings through good housekeeping prior to commencing an investment programme will ensure that investment is optimised and that plant is not over-sized. In addition achieving savings at little or no cost will improve management and employee confidence in the whole process and ease access to funds for future investment.

Project cycle

Implement effective monitoring and targeting

Effective monitoring and targeting (M&T) is the foundation of any energy management programme. The system should be:

- easy to use;
- able to provide a variety of reports;
- able to provide reports to several levels, including end-users, departmental/cost centre managers, site managers, divisional managers (if appropriate) and senior managers.

For more information on M&T see Box 4.3 in the enabling tools section at the end of this chapter.

Identifying projects

Identifying projects requires a combination of technical know-how, which can come from inside or outside the organisation through the use of consultants, equipment vendors or energy service companies and creativity. There is a tendency to produce the same old approaches and the power of creative techniques to re-look at a situation should not be dismissed. The typical tools used to aid in the creation of projects are M&T, Pareto analysis, measurement and verification (M&V), creativity, systems thinking and motivation of end-users. Motivation of end-users is important because ideas for energy reduction may come from anywhere: specialists do not have a monopoly on good ideas.

Developing projects

The development of projects is often overlooked, both in discussions about energy management and in government agency policy. Identifying a concept is not the same as having a viable project on which an investment decision can be made. To move from the concept to a viable project it is necessary to carry out development work which may be in the form of a feasibility study or due diligence. The tools most applicable to this stage of the process are creativity, engineering skills, financial analysis skills, risk analysis modelling skills and systems thinking. Risk analysis is often neglected in development of energy efficiency projects and indeed energy efficiency investments are often labelled as 'low or no risk' without any supporting analysis. Box 4.2 details an approach that takes into account the various project risks.

Too much effort is often spent on identifying concepts, that is, in identifying a potential for energy (or carbon) saving relative to the development of those concepts to a stage at which an investment decision can be made. This is an endemic problem with 'energy audits' and 'energy surveys', particularly when they are supported or subsidised by government energy efficiency agencies. Most of the so-called 'energy audits' or 'energy surveys' do not go into sufficient detail or analysis to make an investment possible. This is a barrier to further action and is a failing of many national energy efficiency programmes promoted by government agencies. Reporting that there is potential for cost-effective investment to produce 20 per cent savings is not the same as developing the projects to prove that they are truly viable and economic and giving sufficient information to justify an investment by the energy user or an energy service company.

Conventional energy audits, which are usually very short, typically only take a 'snapshot' of the energy situation in a facility and do not take into account many of the variables over time such as production variations, potential change of use and so on. The evidence is that the level of savings that they project are inaccurate (+/–20 per cent) (Hansen and Brown, 2004). To address this issue it is important to distinguish between a short survey or audit, which should be

Box 4.2 The project development process

Many energy efficiency projects are evaluated in a very simplistic way. Capital costs are assessed and energy savings are calculated as a single figure, often based on steady state assumptions about the system in question. For instance, the hours run will dramatically affect the economic return from a piece of energy saving equipment but usually a single, standard figure is used, even though actual hours run is unknown (and in some cases unknowable without extensive, lengthy and expensive monitoring). Likewise the performance of equipment such as boilers and refrigeration equipment is often assessed at design conditions. In practice conditions will vary over time in some way. The uncertainties of energy efficiency investment are often ignored by energy efficiency specialists, but they can be a major barrier to organisations making an investment decision. When faced with an investment decision with an unstated and unknown degree of uncertainty, management often find it easier to not invest.

A development process that addresses the uncertainties is described below.

Following an Opportunities Assessment, the list of possible energy conservation projects is assembled and a mathematical model of the process or building and the proposed energy conservation projects is built. This model can be quite simple but must identify the key assumptions made in assessing the savings such as hours run, operating temperatures etc. Within the model the assumptions are varied to assess the net value of the savings using Monte Carlo simulation, which produces a range of net values. Sensitivity analysis is used to determine the key parameters that most affect net value. These key parameters are then measured using measurement and verification techniques to check their variability. Then, the refined data on the variability is used to produce a refined net value calculation.

This approach recognises the risk factors inherent in assessing any energy efficiency project and deploys measurement and verification resources in a focused way that reduces cost.

labelled an Opportunities Assessment, and a full due diligence exercise that can lead to an investment decision, called an Investment Grade Audit. The Investment Grade Audit is a feasibility study that looks at project viability in detail, examines the economics, looks at the risk factors, and has some kind of sensitivity analysis and a clear implementation route. The output should be a document that can be used to make a clear 'go/no go' investment decision and therefore its form and content should be guided by capital investment decision processes within the organisation. Any organisation wanting to move into investing in energy efficiency projects must set a sufficient budget to fund an Investment Grade Audit (or Audits) on the proposed measures as well as an initial energy survey or Opportunities Assessment. All too often there is an initial survey but no follow-up, leading to frustration amongst energy specialists, lack of understanding by government energy efficiency agencies about why the potential is not being exploited, and lost profit for the organisation. Any ESCO contemplating an investment in a client's site should also utilise the Investment Grade Audit approach.

Implementing projects

Implementation requires traditional project management and engineering skills. Post-implementation commissioning of energy consuming or converting plant is still often neglected, but is a vital area. The author has personal experience of monitoring a 'low energy building' that was consuming twice the design energy usage simply due to poor commissioning of heating controls. Use of building energy management systems (BEMS) and supervisory control and data acquisition (SCADA) systems can aid system commissioning of sophisticated buildings and processes and should be planned for in any project implementation programme.

Monitoring results

Post-investment analysis of energy efficiency projects is quite rare. The reasons for carrying it are to maintain the initial level of savings and to learn what works and what does not. The techniques to be used include monitoring and targeting (M&T) and measurement and verification (M&V).

COMMON FAILURES OF ENERGY MANAGEMENT PROGRAMMES

The following are common failures of energy management programmes (see Figure 4.4) which should be avoided:

- Moving straight to the investment stage prior to optimising energy use with existing plant and equipment.
- Failing to maximise the involvement of all staff.
- Failing to clearly allocate responsibility for energy use to line managers, department heads or building managers. This leads to energy becoming 'something the engineers look after'.
- Failing to establish a clear baseline and normalisation procedure.
- Inadequate M&T (feedback too infrequent, for example, monthly).
- Failure to engage senior management, that is have a strategy and programme backed by senior management.
- Undertaking an energy audit or survey, often subsidised by a government agency, without having the means to proceed to an Investment Grade Audit.

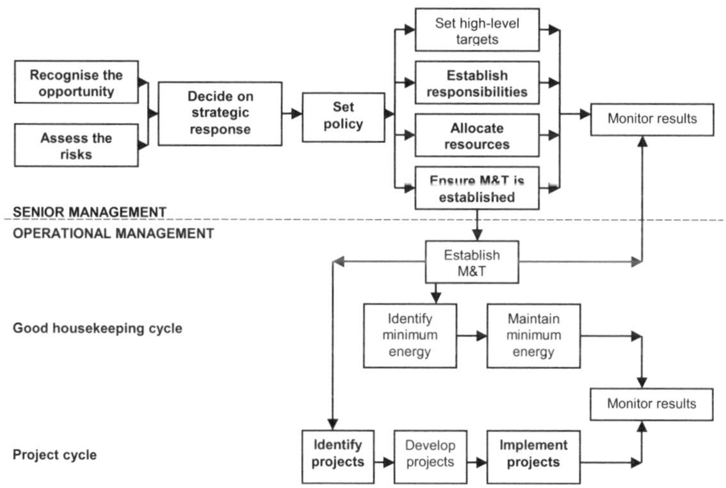

Figure 4.4 The energy demand management process

ENABLING TOOLS

The boxes below describe the basic principles of important enabling tools that are commonly used to support the energy management process. A large body of literature exists on all of them and some references can be found at the back of the book.

Box 4.3 Monitoring and Targeting (M&T)

'Monitoring and targeting' is a cost-saving management technique, designed to detect and diagnose wasteful use of energy, water and other consumable resources. The technique based on the following ideas:

- that consumption volumes are related to measurable driving factors (weather, levels of production outputted, and so on);
- that knowing the driving-factor values it is possible to estimate a target (in the sense of an expected quantity) for each 'stream' of consumption;
- that the onset of waste may be signalled by an unexpected discrepancy between actual and expected consumption;
- that underlying waste may be exposed by the relationship between consumption and its driver(s) being in some way anomalous.

Source: Vilnis Vesma, http://vesma.com/mt/index.htm (accessed 10 December 2006)

M&T is an essential foundation of any energy or carbon management programme and its importance cannot be overstated.

Box 4.4 Measurement and Verification (M&V)

'Measurement and verification' came about as a result of the rise of energy performance contracting in the USA and the need to prove that energy efficiency investments had in fact made savings. M&V has evolved into an international protocol (the International Performance Measurement and Verification Protocol – IPMVP).

Essentially the IPMV defines protocols for measuring energy savings, which vary from using engineering calculations (stipulated) through to actual metering of projects. The choice of M&V technique varies from case to case: a relatively low-cost investment in replacement lighting, for example, will require less actual metering than a complex and more expensive heat recovery installation. The savings for the lighting example may be determined reasonably accurately from engineering calculations using the published efficiencies of the existing and new lighting systems and the hours of use.

Source: www.evo-world.org

The IPMV is still not widely used in the UK but its application would help build confidence in energy savings projects. M&V should be built in to any significant energy or carbon saving project.

Box 4.5 The energy management matrix

The energy management matrix was developed in the 1990s as a technique for assessing an organisation's position in the different areas important for a successful energy management programme. These areas are energy policy, organising, skills and knowledge, information systems, marketing and communications and investment. The matrix is a useful tool for management and external agents of change (consultants or government agencies) to assess an organisation's position and capabilities in energy management. It can be used to determine priorities for improvement and as a tool to measure progress over time.

To use the matrix assess where the organisation is in each of the six categories; for instance, for company *A*, which has:

- no explicit energy policy;
- a part-time energy manager;
- informal contacts between engineer and users;
- cost reporting based on invoice data;
- informal contacts used to promote energy efficiency;
- investment using short-term payback criteria only;

the matrix would look like Figure 4.5.

Company *A* developed an energy management programme using consultants help management implement a programme including:

- an energy policy backed by senior management;
- an energy manager reporting to a committee;

- energy committee used as main channel for contact with users;
- a comprehensive energy monitoring and targeting system;
- awareness training and regular publicity campaigns;
- evaluation of energy saving project using the same criteria as other investments.

Completing a new matrix after the actions (see Figure 4.6) have been taken graphically shows a clear improvement. The matrix can be used to benchmark sites against other sites within a group or externally where data is available.

Level	Energy Policy	Organisation	Motivation	Information Systems	Marketing	Investment
4	Energy policy, action plan & regular review have commitment of top management as part of an environmental strategy.	Energy management fully integrated into management structure. Clear delegation of responsibility for energy consumption.	Formal & informal channels of communication regularly exploited by energy manager and energy staff at all levels.	Comprehensive system sets targets, monitors consumption, identifies faults, quantifies savings and provides budget tracking.	Marketing the value of energy efficiency & the performance of energy management both within the organisation & outside it.	Positive discrimination in favour of "green" schemes with detailed investment appraisal of all new-build & refurbishment opportunities.
3	Formal energy policy, but no active commitment from top management.	Energy manager accountable to energy committee representing all users, chaired by a member of the managing board.	Energy committee used as main channel together with direct contact with major users.	Monitoring & targeting reports for individual premises based on sub-metering but savings not reported effectively to users.	Programme of staff awareness & regular publicity campaigns.	Same pay-back criteria applied as other investments.
2	Un-adopted energy policy set by energy manager or senior departmental manager.	Energy manager in-post, reporting to ad-hoc committee, but line management and authority unclear.	Contact with major users through ad-hoc committee chaired by senior departmental manager.	Monitoring & targeting reports based on supply meter data. Energy unit has ad-hoc involvement in budget setting.	Some ad-hoc awareness training.	Investment using short-term payback criteria only.
1	An unwritten set of guidelines.	Energy management is the part-time responsibility of someone with only limited authority or influence.	Informal contacts between engineer and a few users.	Cost reporting based on invoice data. Engineer compiles reports for internal use within technical department.	Informal contacts used to promote energy efficiency.	Only low-cost measures taken.
0	No explicit policy.	No energy management or any formal delegation of responsibility for energy consumption.	No contact with users.	No information system. No accounting for energy consumption.	No promotion of energy efficiency.	No investment in increasing energy efficiency in premises.

Figure 4.5 Energy management matrix Company *A* before programme

Level	Energy Policy	Organisation	Motivation	Information Systems	Marketing	Investment
4	Energy policy, action plan & regular review have commitment of top management as part of an environmental strategy.	Energy management fully integrated into management structure. Clear delegation of responsibility for energy consumption.	Formal & informal channels of communication regularly exploited by energy manager and energy staff at all levels.	Comprehensive system sets targets, monitors consumption, identifies faults, quantifies savings and provides budget tracking.	Marketing the value of energy efficiency & the performance of energy management both within the organisation & outside it.	Positive discrimination in favour of "green" schemes with detailed investment appraisal of all new-build & refurbishment opportunities.
3	Formal energy policy, but no active commitment from top management.	Energy manager accountable to energy committee representing all users, chaired by a member of the managing board.	Energy committee used as main channel together with direct contact with major users.	Monitoring & targeting reports for individual premises based on sub-metering but savings not reported effectively to users.	Programme of staff awareness & regular publicity campaigns.	Same pay-back criteria applied as other investments.
2	Un-adopted energy policy set by energy manager or senior departmental manager.	Energy manager in-post, reporting to ad-hoc committee, but line management and authority unclear.	Contact with major users through ad-hoc committee chaired by senior departmental manager.	Monitoring & targeting reports based on supply meter data. Energy unit has ad-hoc involvement in budget setting.	Some ad-hoc awareness training.	Investment using short-term payback criteria only.
1	An unwritten set of guidelines.	Energy management is the part-time responsibility of someone with only limited authority or influence.	Informal contacts between engineer and a few users.	Cost reporting based on invoice data. Engineer compiles reports for internal use within technical department.	Informal contacts used to promote energy efficiency.	Only low-cost measures taken.
0	No explicit policy.	No energy management or any formal delegation of responsibility for energy consumption.	No contact with users.	No information system. No accounting for energy consumption.	No promotion of energy efficiency.	No investment in increasing energy efficiency in premises.

Figure 4.6 Energy management matrix Company *A* after programme

SUMMARY

The processes and tools required to implement and maintain an effective energy management programme have been neglected during the period of low energy prices and there is an urgent need in many organisations to reinvigorate energy management. Using the processes and tools described above, many organisations should be able to start a programme that will save anywhere between 10 and 30 per cent of energy use, possibly more through the application of techniques such as whole system design.

Box 4.6 Energy benchmarking

Benchmarking can be defined as 'the process of comparing a product, service process – indeed, any activity or object – with other samples from a peer group, with a view to identifying "best buy" or "best practice" and targeting oneself to emulate it' (Williams, 1994).

'It is common for benchmarking to be incorrectly mixed up with performance measurement techniques' (McDougall and Hinks, 2000).

Several industry sectors have developed energy efficiency benchmarking standards. The longest standing of these in the UK is probably the system used in the brewing industry, which has been operating since the early 1980s. The scheme is based on biannual surveys of the entire brewing industry and the overall measure of performance is MJ/hl (megajoules of energy used per hectolitre of beer output). The scheme has in-built correction factors for differing size of brewery and differing percentages of packaging such as 'bulk' in kegs and 'small pack' bottles and tins.

Other UK trade associations operate similar schemes and these schemes are the basis of measuring progress in the Climate Change Levy Negotiated Agreements.

In the local authority building sector the UK Audit Commission (the independent body responsible for ensuring that public money is spent economically, efficiently and effectively in the areas of local government, housing, health, criminal justice, and fire and rescue services) developed the Normalised Performance Indicator (NPI) system of benchmarking building energy use.

The NPI system took into account building type (for example school, leisure centre, elderly people's homes, etc.), local weather through degree days, and location in terms of exposure (exposed, normal or sheltered). The NPIs were based purely on a comprehensive survey of many thousands of local authority buildings. They did not take into account any inherent features of the building age, design or construction type, but merely recorded the distribution of energy use for each category of buildings. It then allowed each authority to compare the performance of buildings on a reasonably consistent way and to measure its own improvement as an energy management programme was implemented.

The NPI system was subsequently expanded to many other building types outside local authorities. Some agencies made the mistake of converting these to £/m^2. It is essential to maintain the NPIs in energy units due to the fluctuation of energy prices.

The use of benchmarking is particularly helpful in assessing overall performance and overall potential for saving. It should, however, be used with caution. For instance, two buildings in the same category, building *A* and building *B*, may have NPIs of 140 and 120 respectively. This could mean that building *A* has a larger potential to save; but this may not be true if, for example, building *A* is an old building with a structure that cannot be upgraded in an economic way (for example, solid brick). Building *B* may have a greater potential for cost-effective savings, despite having a better NPI.

Box 4.7 Opportunities Assessment (energy audit or energy survey)

A common component of energy management programmes has been 'energy audits' or 'energy surveys'. Even now, the first call from a company looking at energy costs is often for an energy survey. Most of these would be better labelled as Opportunities Assessments as they are not rigorous technical or economic studies or audits but rather simply identification of potential opportunities which should be evaluated further. They have a place at the beginning of an energy management programme and possibly at regular, but infrequent intervals (three to five years or after any major change to a plant or a building) but they should not be regarded as an end in themselves.

Research has shown that these energy surveys have an error of +/–20 per cent in their estimates of saving (Hansen & Brown, 2004). A typical energy survey is a 'snapshot' that relies on steady state assumptions and ignores changes in many variables such as occupancy, hours of use and operating temperatures.

In many energy surveys M&T and 'good housekeeping' is often arbitrarily reported as 5 per cent savings irrespective of any other factors and without any evidence. This is just a convenience used by low-quality consultants to boost savings potential figures.

Government agencies in some countries have used the results of these assessments to proclaim publicly on the potential for energy saving without taking into account their inaccuracies and the fact that many of the opportunities identified are not practical to implement, not technically appropriate or not economic.

Box 4.8 Investment Grade Audit (due diligence/feasibility study)

The problems with energy audits identified above have led some experts to coin a new phrase, the Investment Grade Audit (IGA). This terminology is extremely useful as it separates the activity from energy audits or surveys and focuses on the true purpose of the work, to deliver the information needed to make an investment decision. Other terms used to describe this activity include 'due diligence' and 'feasibility study'.

The IGA can be focused on a particular technical measure or on all identified opportunities at a particular site. It should examine all the site-specific technical factors that affect the viability and economics of the measure(s) being looked at. It will often require on-site measurement of parameters such as temperatures and pressures or the operation of plant and machinery, that is M&V. The main difference between an IGA and an Opportunities Assessment is that the Opportunities Assessment can only look at a single snapshot in time, whereas the IGA needs to consider how the parameters of the system vary over time: daily, weekly and monthly. The use of M&V, along with the time involved, means that an IGA will cost considerably more than an energy survey (between 5 and 10 per cent of energy spend or the specific project capital investment). Depending on the situation and the size and complexity of the investments being considered it will take between 10 and 100 man-days of work to complete.

Box 4.9 Creativity techniques

To identify or create investment opportunities requires information about energy usage and costs that can come from an M&T system coupled with a knowledge of relevant technical concepts and an often ignored component: creativity. Creativity may redefine what is relevant, and the use of creativity techniques may be helpful at the opportunity definition stage where a creative synthesis of internal and external information is necessary.

There are many creative techniques available – (over 200 according to www.mycoted.com, which lists an A to Z of them). The best known, all too often used loosely and incorrectly, is brainstorming. Others that may be useful in an energy context include alternative scenarios, assumption surfacing, gap analysis, lateral thinking, synectics, technology monitoring and value engineering.

An external consultant, or an outsource service provider, can act as a 'gatekeeper' or perhaps more accurately defined 'a hunter gatherer', bringing in external ideas and expertise to the process. It should be noted, however, that many energy engineering consultants will tend to bring the conventional ideas and approaches.

Non-technical staff can sometimes be more creative about identifying energy saving opportunities than technical staff, possibly because they are not overloaded with the knowledge of what is not possible.

Box 4.10 Holistic systems design

Whole-system design optimises an entire system to capture synergies. The concept is straightforward, but implementation is not trivial. It requires creativity, good communication and a desire to look at causes of problems rather than adopting familiar solutions – and it requires getting to the root of the problem: education. Like the engineering profession itself, engineering education is often compartmentalised, with minimal consideration of systems, design, sustainability and economics. It stresses analysis over synthesis. The traditional engineering design process focuses on optimising components for single benefits rather than whole systems for multiple benefits – thereby 'pessimising' the system. This, plus schedule-driven 'repetitis' (copying the previous design), perpetuates inferior design. Whole-system design, on the other hand, offers competitive advantage by revealing better, simpler solutions.

Whole-system design has already proved its value in industrial engineering. More than half the world's electricity turns electric motors. The largest use of electric motors is pumping. In 1997, a major carpet manufacturer was building a factory in Shanghai. One heat-transfer loop was designed to use 14 pumps totalling 95 horsepower. Using whole-system design that RMI's Amory Lovins brought from Lee Eng Lock in Singapore, Dutch engineer Jan Schilham cut the power use by 92 per cent to just 7 horsepower by using fat, short, straight pipes rather than skinny, long, crooked pipes. Thanks to smaller motors and pumps, total capital cost went down.

Source: www.rmi.org

Box 4.11 Project management

Project management uses a range of tools to optimise the use of resources, time, money and people over the duration of a project. Project management is about controlling five variables:

- scope
- risk
- quality
- cost
- time

– although often the focus is on cost and time. Delivering energy efficiency projects, like any other projects, requires good project management.

Box 4.12 Motivation and awareness training

Motivation and training is often over looked in energy management programmes. As discussed above it is important to maximise savings through good housekeeping before launching an investment programme. This is best done by motivation and awareness training.

Successful motivation and awareness training programmes include the following elements:

- training
- incentives
- feedback.

Without all of these elements the programme's effectiveness will be severely reduced. Staff generally respond to:

- linking the organisation's energy use to global issues such as climate change;
- training that gives them specific advice on how to reduce energy use;
- incentives such as projects to improve the working or local environment;
- clear, understandable feedback on a regular basis that allows them to take effective action.

Box 4.13 Investment analysis

Any proposed energy efficiency project should be subjected to investment analysis. The complexity of this analysis will depend on the project costs and risks. A small, low-risk project may only require a simple payback calculation to justify whereas a proposed investment in a major piece of energy infrastructure, for example a combined heat and power (CHP) plant, will require complex financial modelling leading to a discounted cash flow calculation. Anyone promoting the adoption of energy efficiency projects needs to be familiar with the techniques of investment analysis, such as internal rate of return (IRR) and net present value (NPV), which are well documented in textbooks.

Box 4.14 Carbon management

Carbon management is a tool developed by the Carbon Trust for reducing carbon intensity. It is a systematic process that addresses all carbon emissions resulting from an organisation's activities and can include transport and supply chains, thus moving beyond the normal boundary of energy management.

The process has five steps:

- Identifying internal stakeholders and gaining their buy-in and agreeing the programme and responsibilities and commitments.
- Evaluating the business case: this involves developing and agreeing an overall understanding of climate change and the risks and opportunities it presents, establishing a carbon baseline, identifying carbon reduction drivers and appraising value at stake.
- Identifying opportunities that require identifying company-wide emissions assets and liabilities and producing and evaluating proposals for carbon emission reduction projects.
- Prioritising ideas and developing a specific, viable implementation plan.
- Implementation, which includes monitoring progress against the implementation plan and its effectiveness in delivering the specified objectives, reviewing progress and updating the plan accordingly.

For more information on carbon management see the Carbon Trust brochure *Carbon Management*. Available from: www.carbontrust.co.uk/publications/publicationdetail?productid=CTC506

Box 4.15 Contextual analysis

Although not a technique in itself, contextual analysis of any proposed energy efficiency or carbon reduction project is an essential part of the investment decision process. Often such investments fail to take into account the impacts of other decisions being planned or made within the organisation. Examples encountered include investing in an energy improvement for production line or an entire facility that has a limited lifetime and investing in a heat recovery system when there is already a plan to change a process for quality reasons that will remove the source of the heat being recovered.

Failure to carry out a contextual analysis will lead to sub-optimal return on investment.

PART 3

Outsourcing Energy Management

Outsourcing is the act of transferring some of an organisation's recurring internal activities and decision rights to outside providers, as set forth in a contract.

Maurice F. Greaver

Don't own a cafeteria: get a food company to do it. Don't run a print shop: Let a printing company do that. It's understanding where your real value added is and putting your best people and resources behind that. Back rooms by definition will never be able to attract your best. We converted ours into someone else's front room and insisted on getting their best. That worked for us so many times. This is what outsourcing is all about.

Jack Welch

CHAPTER 5

Why Outsource?

INTRODUCTION

Outsourcing of all types is growing rapidly for the following reasons:

- Competitive pressures are becoming more severe in the global economy. Most products, including those that once relied on a craft approach, such as beer, can increasingly be made anywhere in the world. Most multinational companies are moving production towards lowest production cost sites, wherever they may be in the world and within manufacturing organisations it is now normal for new production projects to be tendered out to different production sites. In some industries whole elements of production, such as packaging, can be outsourced to other companies. In some cases, for example, in the brewing industry, companies that provide outsourced services such as packaging can even be competitors to the client company in other markets.
- Product and cycle times have been reduced dramatically (therefore there is a need for a quicker response and more flexibility in the supply chain including energy supply systems).
- Investor demands have focused management on delivering results.
- Small agile niche competitors can change industries and cost structures over night.
- Achieving significant operating and financial performance improvements is critical (the twin pressures of oil peaking leading to sharp rises in energy prices and global climate change described in Part I mean that significant improvements in efficiency are needed now, just to stand still. Incremental change is no longer good enough).

An added motivation to outsourcing including asset monetisation can be the method of measuring success in a company. If a company uses return on capital employed (ROCE) or added value to measure success then there is a motivation to reduce the amount of capital employed. This will increase the motivation to sell assets to get them off the balance sheet and hence outsourcing involving asset monetisation will be more attractive. (This kind of off-balance-sheet treatment is a complex issue and in every case it is necessary to seek appropriate legal and accounting advice at an early stage, especially in the post-Enron world.)

Whatever the size of the organisation, the processes of energy management (described in Chapter 4) are the key to achieving real energy cost savings and energy risk management. The resources dedicated to energy management will vary between organisations but the basic process steps are the same. So why should an organisation consider outsourcing some or all of its energy management?

Organisations are facing the following energy problems:

- Increasing energy costs (and increasing price volatility).
- Ageing utility infrastructure (in many instances companies have repeatedly invested in new process equipment without updating utility equipment).
- Lack of resources. In the 1980s large organisations often had dedicated energy management teams but most of these were disbanded when energy prices were low. Organisations downsized and lost energy management skills during the period of low and stable energy prices and expertise was lost from the economy as a whole as people moved into other functional areas. Some organisations are now (as of 2006) appointing energy managers to deal with the energy issue but the supply of qualified, experienced people is a problem.
- Pressure to focus on core business and outsource non-core activities. This is a general business pressure and there has been a rapid growth in outsourcing of non-core activities over the last 20 years.
- The increasing pressure to reduce emissions to mitigate climate change.
- Increasing complexity of the regulatory environment bought about by the onset of schemes such as Integrated Pollution Prevention and Control (IPPC), Climate Change Levy (CCL) and EU Emissions Trading Scheme (EU ETS).

The decision to outsource energy management can be driven by several factors, which may include the setting of a high-level target to reduce energy and utility costs and/or emissions; a threat of plant closure, which triggers a fundamental examination of costs and evaluation of options that may previously have been considered 'off limits'; or the recognition of the need to invest in new utility plant at a time when capital is constrained due to other investment priorities.

According to Greaver (1999) some reasons to outsource any service or process include:

- obtaining expertise, skills, best practice and technologies that are not available within the organisation;
- increasing flexibility to meet changing business conditions, demand for products/services and technologies;
- improving operating performance;
- reducing costs through superior provider performance and the provider's lower cost structure;
- enhancing effectiveness by focusing on core business;
- transforming the organisation;
- reducing investment in assets, freeing up resources for other purposes;
- improving risk management;
- gaining market access and business opportunities through the provider's network;
- improving management and control;
- improving credibility and image by associating with superior providers (gain new assets that improve credibility/image);
- expanding sales and production capacity during periods when such expansion could not be financed;
- turning fixed costs into variable costs;
- commercially exploiting existing skills;
- acquiring innovative ideas;
- giving employees a stronger career path;
- increasing commitment and energy in non-core (but important) areas;
- generating cash by transferring assets to the provider;

- accelerating expansion by tapping into the providers resources, capacity, systems and processes;
- increasing product and service value, customer satisfaction and shareholder value.

Examples of these reasons in the context of energy management are shown in Table 5.1.

BARRIERS TO THE OUTSOURCING PROCESS

Factors that often inhibit organisations from outsourcing in general include:

- uncertainty;
- fear of:
  - loss of control;
  - loss of core competencies;
  - employee dissent or unrest;
  - difficulty in reversing decision;
  - risks of failure.

Another issue hindering the take-up of advanced energy services is the energy procurement process. The annual power and gas procurement round, in which companies tender the supply of power and gas but focus entirely on the price per MWh, may achieve the lowest unit cost for commodity but is a barrier to obtaining the lowest overall cost of delivered energy. There is a need to break this process and look at longer-term deals tied with energy services aimed at giving lowest total costs. Senior managers need to change the established methods of energy procurement and focus on total cost, not just commodity cost. Energy procurement teams will always say they have the best deals; but have they considered the risk issues for instance? Have they really optimised energy volume, energy price and energy risks?

Another, related factor that inhibits outsourcing of energy functions is that the various elements of total energy cost sit in different budgets or functional areas within most organisations in areas including maintenance, capital projects, energy procurement, bill verification and accounting, monitoring and targeting (M&T) and management reporting. A major inhibitor is that many organisations still see energy as an uncontrollable overhead rather than a management activity that can be used to produce value and manage risks.

Table 5.1 Reasons to outsource energy management

Reason to outsource	Energy or utilities example
Obtain expertise, skills, best practice and technologies that are not available within the organisation	This is a particularly important in energy management. In the 1970s and early 1980s many organisations had effective, large energy management groups. With the lower energy prices brought about by the advent of competitive energy markets, coupled with the general trend towards downsizing in the 1990s, many organisations reduced, or even lost their energy management teams altogether. The economy as a whole lost expertise in energy management. With the increase in energy prices in 2004/05 many organisations did not have the expertise or skills needed to address the problem and there is a need to reacquire lost skills as well as new skills required by the growth of the carbon agenda.
Improve operating performance	As well as the obvious example of reducing energy costs there are examples where improving the reliability or service levels of utility provision can improve product quality and/or production level, thus improving profitability.
Reduce costs through superior provider performance and the provider's lower cost structure	As well as reducing energy demand outsourcing energy management can potentially access better buying power for utility supplies such as power and gas, or ancillary supplies such as chemicals or plant and equipment, thus reducing costs.
Enhance effectiveness by focussing on core business	By transferring operational problems to a service provider some key staff for example in engineering, may be freed up to focus more on production processes and problems rather than the utility area
Transform the organisation	Outsourcing energy management can bring in new ideas, processes and investment that contribute to transforming the organisation.
Reduce investment in assets, freeing up resources for other purposes	Using an energy service provider to invest in utility projects can liberate investment capital for other uses such as investment in new production equipment or marketing.
Improve risk management	A major benefit of outsourcing energy services is transferring some of the energy related risks to a service provider.
Gain market access and business opportunities through the provider's network	Through energy outsourcing with a utility owned energy service provider a customer could gain access to the expertise needed to package and sell carbon emission credits gained through participation in schemes such as the EU Emissions Trading Scheme.

Table 5.1 ***Continued***

Improve management and control	Utilities may be poorly controlled due to lack of resources and management. Often service levels of internally managed utility operations are not be defined and even considering outsourcing can lead to a clearer definition of service levels.
Gain market access and business opportunities through the provider's network	There may be strategic deals that can be done in which a customer accesses the providers marketing networks. An example may involve the customer utilising a utility-owned energy service company's marketing database to sell its products to the utility's domestic customers.
Improve credibility and image by associating with superior providers (gain new assets that improve credibility/image)	This may be appropriate for smaller and mid-size companies who could gain profile, credibility and image from entering into a partnership with a large, utility-backed energy service provider. More directly of course they could gain new utility assets which help reduce costs and assist in creating a good image to their customers.
Expand sales and production capacity during periods when such expansion could not be financed	An example of this would be entering into an energy services contract in which the provider supplies capital for utility infrastructure investment which is needed to increase system capacity in order to support and increase production throughput. This may be appropriate for both smaller, growing companies where capital is tight, and for larger organisations strategically moving out of direct involvement in energy issues.
Turn fixed costs into variable costs	Total utility costs are made up of fixed costs for example labour and variable costs such as energy input. Some forms of utility outsourcing can switch fixed costs into variable costs for example if utilities are sold at a unit rate covering all costs such as labour and capital. This does, however, involve some form of risk transfer to the service provider who will need to be protected by 'minimum take' clauses. These ensure a minimum level of revenue irrespective of the volume of utility consumed.
Commercially exploit existing skills	A customer may have skills in-house that can be exploited by a joint venture with an energy service company in new areas.
Acquire innovative ideas	A company which is focusing on its core business is unlikely to create, or adapt innovative ideas in the energy/utility area. Although it is unlikely to be an innovator in the sense of developing new technologies, an energy service company can bring with it the ability to spot new opportunities and package existing technologies in a new way for example effluent treatment plants optimised for bio-gas production and then using bio-gas as an energy source.

Table 5.1 ***Continued***

Give employees a stronger career path	Utility employees in most companies have a limited career path unless they move out of the utilities area. For those people who are interested in and committed to a career in this area, the 'glass ceiling' effect can be frustrating. By joining an energy service provider, particularly of it is part of a large O&M/FM or utility group, can open up new career paths. Often training, other than mandatory training on issues such as health and safety or electrical systems, will not be freely available to utilities staff and any budgets for training are cut to the minimum. In a dedicated utility or energy service provider, additional training options will be available. RWE, for instance offers all operational staff mandatory training, professional development training and aspirational training. In several cases staff who were apparently 'stuck' at an operational level have moved onto other roles. Even if promotion is not possible or available, being part of a large energy service company will offer opportunities for re-location or experience within different industrial sectors.
Increase commitment and energy in non-core (but important) areas	Energy is often seen as being of low importance and therefore it becomes of low importance. A service provider can re-focus attention on it without taking away resources from the organisation's core mission.
Generate cash by transferring assets to the provider	Selling utility assets to the service provider will generate cash on the first day of the outsourcing project. This, combined with on-going savings, can bring about a highly cash positive position on the whole outsourcing contract.
Accelerate expansion by tapping into the providers resources, capacity, systems and processes	As well as freeing up capital for investment in the core business there may be opportunities to use a suppliers marketing and customer database to promote the customer's product directly to a new or different market.
Increase product and service value, customer satisfaction and shareholder value	By reducing total utility costs outsourcing can increase the value of a product and service. By reducing environmental impact through reduced energy use, customer satisfaction can be enhanced through their beliefs or perceptions that reduced environmental impact is a positive feature.

Another barrier is the belief that engineering staff have already optimised the process or building(s). This view has to be seriously questioned, especially in the current period when we have experienced a large increase in energy prices after a long period of relatively low prices, which has to mean that there is new potential for cost-effective projects. Also in-house engineers, as well as consulting engineers of most varieties, will tend to design systems in the way they have always been designed: inherent conservatism is built in to the engineering design and specification process. Have they really considered 'information-based engineering sizing' and 'holistic design' (see Chapter 8) in designing new process plant, or have they just optimised individual system components? Have they really considered alternative, possibly radical alternatives, such as on-site renewable generation?

Negativity and lack of credibility can be a major problem. This is characterised by comments such as 'management will never follow through on this' or 'they talk about it but we can never get the investment capital' or 'we tried all this five/ten/fifteen years ago and it didn't work then'. Senior management have to work hard to demonstrate that they are serious in any energy management programme, whether it be in house or outsourced.

In many large, multinational organisations there is an inherent tension between central, head office support staff and on-site operational people. Central staff, particularly in functions such as energy buying, which in many companies is now organised on a pan-European basis, are keen to demonstrate their effectiveness, sometimes by seeking to buy pan-European energy and energy efficiency deals. The reality from the supplier's side is that there is no one who can currently offer this kind of integrated pan-European deal. Some pan-European utilities come close but the differing regulatory frameworks make a unified approach on power and gas sales difficult. Furthermore, the central HQ staff have not always engaged the individual sites sufficiently to make a programme happen. An outsourcing programme driven entirely by the centre but without support from individual sites is not going to move forward. The sites concerned have to be fully engaged and motivated. At the end of the day production units are the source of profits and hence usually the real power to get things moving.

Another major barrier is the view that 'we can't have outsiders touching the process'. This is clearly a real concern that needs to be addressed by a clear definition of roles and responsibilities plus careful selection of an energy partner. It becomes a more valid concern as the energy intensity and the linkages between process and energy use increase, for example, in industries such as glass, cement and chemicals.

The decision to outsource any activity should always be strategically driven. Energy outsourcing has most often been undertaken in a tactical way, aimed at achieving tactical results, and this can lead to under-performance, disappointment and total failure.

CHAPTER 6

The Outsourcing Cycle

This section describes the process of outsourcing from the decision to consider an outsourced solution through to the management of a live contract. At the highest level the outsourcing cycle can be divided into three main steps:

- deciding to outsource;
- procuring a supplier;
- managing the contract.

The three stages of decision, procurement and contract management described in the introduction can be further sub-divided.

Deciding to outsource includes the following sub-stages:

- planning;
- strategic analysis;
- cost and performance analysis;
- risk analysis.

Procurement includes:

- selecting providers;
- due diligence;
- negotiating terms.

Contract management includes:

- transition (prior to services commencement);
- services commencement;
- an extended transition (if required, after services commencement);
- an investment stage (or a learning stage if no investment needed);
- 'normal' operations stage;
- termination.

The major steps and events are shown in Figure 6.1. Each step, along with the sub-stages, will be considered in turn in this section. The question of the scope of outsourcing is also considered. Table 6.1 shows the typical scope of various energy management outsourcing forms.

Table 6.1 Typical scope of various energy management outsourcing forms

Outsourcing form	Activities included in scope
Strategic energy consulting including: • Benchmarking • Opportunity assessment • Policy assistance • Design programme	Recognise the opportunity Assess the risks Decide on strategic response Set policy Set high-level targets Establish responsibilities Allocate resources Ensure M&T is established Monitor results
M&T bureau	Establish and run M&T system
Consulting	Provision of M&T information Motivation and awareness training Identify projects Develop projects Manage implementation of projects
Energy procurement consultancy	Tendering energy supply, market advice Invoice validation
Build own and operate	Develop projects Implement projects Operation and maintenance Finance projects
Utility system operation and maintenance	Operation and maintenance of utility systems
Contract energy management	Operation and maintenance Provide M&T information Identify and develop projects Finance projects (usually supply-side for example boilers and CHP)
Energy savings performance contracting	Identify, develop and implement projects Finance projects or source finance
Utility outsourcing	Identify, develop and implement projects Finance projects Operation and maintenance

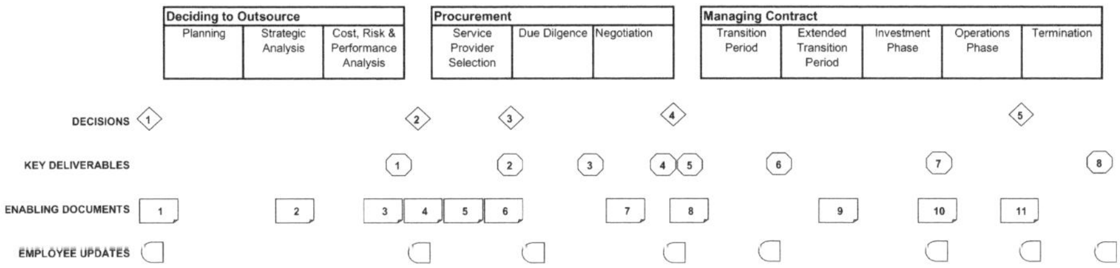

DECISIONS
1 Decision to investigate outsourcing
2 Decision to outsource
3 Decision to proceed with selected preferred supplier
4 Decision to sign final agreements
5 Decision to terminate contract

KEY DELIVERABLES
1 Recommendation to outsource or not
2 Selection of preferred service provider
3 Completed due diligence process
4 Agreed contracts
5 Agreement of Transition Plan
6 Services Commencement
7 Completed investment phase
8 Cleanly terminated contract

ENABLING DOCUMENTS
1 Timeline
2 Strategic analysis
3 Cost/risk and performance analysis
4 Request for Qualification (RFQ)
5 Request for Proposals (RFP)
6 Tender evaluation document
7 Punch lists to monitor negotiation progress
8 Transition plan
9 Investment plan
10 Continuous Improvement plan
11 Termination plan

Figure 6.1 The outsourcing process

It is necessary to say at this stage that the general process described here applies to all forms of energy management outsourcing but some of the sub-stages are more appropriate to particular forms of outsourcing. The descriptions below assume the more complex type of utility outsourcing on the basis that if something is not needed in other types of projects it can be ignored.

PHASE ONE – DECIDING TO OUTSOURCE

PLANNING

A plan covering the process of evaluating outsourcing, as well as an outline plan for the process if outsourcing goes ahead, should be developed with clear milestones and timelines (see Figure 6.1 above). Clearly there is no need to include detailed planning for later stages such as implementation prior to the final decision to outsource but an overall, sensible timescale should be established at the outset. Major outsourcing projects are complex and time consuming, moving from the first idea through to services commencement can take six to eighteen months depending on the size and scale of the contract, the scope of the services being outsourced and the culture of the client organisation. A sample timeline is shown in Figure 6.2.

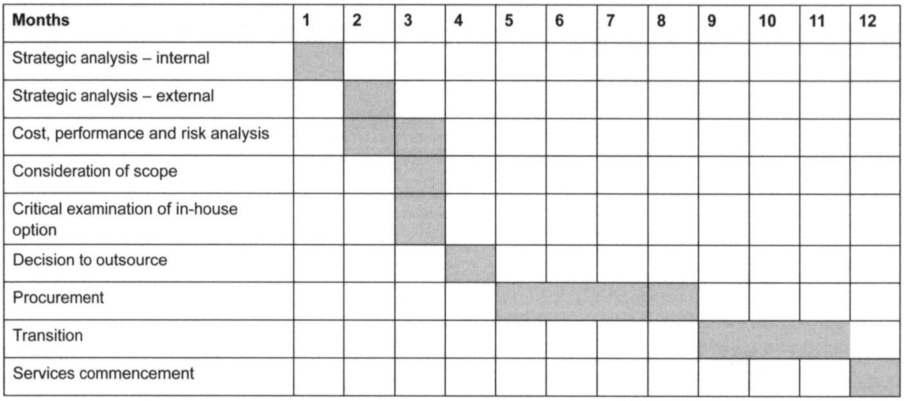

Figure 6.2 Outline timeline for outsourcing

The project needs to have a senior sponsor, probably a board member who should define the terms of reference and provide overall management and leadership. It is important to clearly define a team to work on the project with a team leader responsible for delivering the required outputs. It is likely in most organisations that the team members will have other everyday responsibilities, although if possible they should be seconded to the team and relieved of everyday work. An ideal team would include technical, financial and commercial people, as well as any specialised resource such as legal input as required. Other resources needed for the evaluation phase, for example, an external outsourcing consultant, need to be defined and agreed. A specialised consultant can be extremely useful, particularly for organisations who have not outsourced before, although the supply of consultants experienced in this field is extremely limited.

The deliverables from the first phase are:

- a detailed plan;
- a strategic analysis;
- a cost, performance and risk analysis.

The content of these are examined in turn below.

STRATEGIC ANALYSIS

The strategic analysis for energy outsourcing should have an external and an internal element:

- future energy scenarios (external);
- whether or not energy is a core competence for the organisation (internal).

Consider future energy scenarios

Clearly an important factor affecting a company's strategy is its view of the energy future. Part I outlined the major issues facing the world in the energy arena, namely the need to make the transition to a low carbon economy and the threat of oil peaking. It is clear that the energy future will be different to the present and the past, and to make good decisions about an individual company's energy future it is essential to understand the dynamics of change in the energy field itself. There will be differences in emphasis between companies, with some companies embracing a more radical environmental agenda than others; BP, for instance, has engaged in the need to build a low carbon economy and led by example in installing photovoltaics and local wind generators in its petrol stations. Other oil companies have not taken such an assertive stand. Companies in other sectors, especially retailers like Sainsbury's have invested heavily in energy efficiency measures, whilst in the automotive manufacturing sector Ford in the USA have invested in greening the Red River plant in a radical way (including grass roofs and reed bed effluent treatment systems: see McDonough and Braungart, 2002), and in the UK installed embedded wind turbines at the Dagenham site. Other companies will choose a less radical approach, focusing purely on cost management. As stated earlier however, the decision to 'carry on as normal', be it taken explicitly and consciously – or more likely implicitly and unconsciously, in itself carries considerable risks. If you take this decision for your company and then oil peaking leads to oil prices reaching and sustaining record highs for many years, you may be out of business. Likewise, a world in which carbon emissions become more restrained may threaten your business through cost penalties, fines and lost reputation.

In considering energy futures work such as Shell's scenario planning may be helpful (Shell, 2005).

Consider whether energy is a core competence

There are several definitions of core competencies, including the following:

> *[T]he innovative combinations of knowledge, special skills, proprietary technologies, information, and unique operating methods that provide the product or service that customers value and want to buy.*
>
> (Greaver, 1999)

> *A core competence is a combination of complementary skills and knowledge bases embedded in a group or team that results in the ability to execute one or more critical processes to a world-class standard.*
>
> (Coyne, Hall and Clifford, 1997)

Identifying core competences is not a simple exercise but requires extensive critical examination of the root causes of product or service attributes and is part of a larger management issue than outsourcing. For most organisations, other than perhaps very energy intensive manufacturing such as glass and cement, it is hard to see energy and the operation of energy systems as being a core competence. It is a critical competence, in that if it fails to deliver production will stop, but it is not really core to most organisations. This is a major driver for energy management outsourcing.

COST, PERFORMANCE AND RISK ANALYSIS

The cost, performance and risk analysis should:

- summarise the current and projected costs, performance and risks;
- benchmark these where possible against competitors and companies operating in other comparable sectors;
- identify which risks are currently being taken and the view on the probability and impact of adverse events.

The analysis should consider future costs over a five- to ten-year period (with appropriate sensitivity analysis) including foreseeable capital requirements, for example known plant replacement requirements such as a new boiler house. This analysis should be robust enough to form the baseline to which any outsourcing proposal is compared.

Assess current and future utility costs

Data on the elements of utility cost needed to assess current and future utility costs should be available from management accounting systems and energy monitoring and targeting (M&T) systems. Elements to include when considering total costs are:

- energy
- operations and maintenance (O&M) labour
- O&M
- capital depreciation.

A 'base case' model of future costs needs to be made and it is this model that will form the basis of cost comparison for the evaluation of bids from outsourcing vendors. The model should consider the total cost of ownership of the complete energy system. The assumptions behind the model need to be agreed and signed off by relevant members of the project team, both technical and financial. The model should be able to take into account projections of future energy prices (the model could include different scenarios for future energy prices to assess their impact), forecasts of future labour costs and foreseen capital investment needs. The latter may be difficult if there is no existing long-term capital replacement plan, but in this case the project team will have to take a hard look at the age and condition of equipment, its design life and likely replacement costs, all of which may require some input from specialists or suppliers.

As well as the cost model the following factors should be assessed:

- Overall energy efficiency (benchmarked against the international industry where possible).
- Overall labour productivity (number of utility staff per energy spend). Data on this may be difficult to acquire but it should be possible to compare with other similar sites within a group or outside companies. Many companies other than direct competitors are usually open to exchange of such data.
- Overall quality of utility service – this should be quantitative if possible.
- The improvement of quality achieved over the last year, three years and five years.
- The improvement in energy efficiency achieved over the last year, three years and five years.
- The usefulness of energy information systems. How useful are the existing energy information systems? Are they actually being used or are reports simply being produced and distributed because they always have been?
- Outputs of each activity.
- Costs at an activity level.

Where possible these parameters should be benchmarked, both within the company if it has other production sites or buildings and within the domestic industry, and internationally.

It is important to estimate the cost of poor performance, that is, the cost of doing nothing. This will include future costs as energy prices increase as well as the cost of future replacement needs. The costs of outsourcing, including one-off cash flows such as the cost of training, legal costs and the costs associated with the transfer of staff, need to be added to the model described above to fully evaluate the outsourcing option.

Risk analysis

The project team should identify the risks the company is taking in the utility area. As these risks have, in most cases, been taken for many years they are often not explicitly stated.

This section sets out the various risks associated with the use and management of energy. An understanding of the risks is crucial to energy management and satisfactory outsourcing of energy management activities. The risks to be considered in different categories include:

Commodity-related risks

- commodity price
- security of supply
- quality

Conversion-related risks

- efficiency risk
- demand risk
- correlation risk

Operations- and maintenance-related risks

- risk of failure leading to repair
- risk of failure leading to replacement

Project-related risks

- construction risk (time and schedule)
- performance risk

Environmental and regulatory-related risks

- Climate Change Levy risk
- EU Emissions Trading Scheme risk
- Integrated Pollution Prevention and Control risk

Brand- and image-related risks

- risk of an environmental incident
- risk of being perceived as environmentally damaging through normal operations

These risks are now considered in turn.

Commodity-related risks

Price risk

Price risk is the risk of variations (usually upwards) of energy prices. Traditionally these have not been managed at all and as was seen in the section on energy supply management most organisations purchased energy on an annual basis, going out to tender for their energy use on a single day. Once the price was fixed and the contract signed the price risk was 100 per cent fixed (hedged) for one year and one year only. This may have been acceptable when prices were relatively low and stable but is no longer acceptable to many organisations as energy prices are high and very volatile.

In the early days of energy performance contracting energy service companies in the United States sometimes took energy price risk as part of 'shared savings' energy performance contracts which was a risk that they had no way of managing or controlling. This led to several failed energy saving performance contracts and subsequent court cases that gave energy service companies a bad name.

The new breed of energy procurement strategies referred to in Chapter 4 offer new ways of managing the price risk. Different approaches are possible but essentially they involve having close and constant contact with the energy market, a transparent deal with an energy supplier and buying different parts of the total energy demand at different times, for example, buying a year's worth of base load and then buying peaks at different times according to variations in prices. With these procurement methods it is necessary for the organisation to take a view on its approach to price risk. Different companies will have a different risk profile – a publicly quoted company, for example, may value a lower-risk profile with greater price certainty than a privately held company which is more willing to take a higher risk in return for a lower price with greater volatility.

An energy service company (ESCO) who works closely with or is part of an energy supplier can access tools that allow the client to manage the risks more effectively. This is one of the advantages of working with an ESCO linked to an energy supplier.

Security of supply

Clearly, whatever your product, be it based on heavy engineering or intellectual property, you will not be able to produce it without a reliable source of appropriate energy to your process, building or IT. There is always a risk that supply of the basic commodities, power, gas and oil, will be disrupted by failures in the system caused either by technical failures, extreme weather events, industrial action (for example the petrol tanker drivers' strikes in the UK in 2001) or terrorism. For end-users this risk falls into two sub-areas: the risk of failure of energy commodity supplies to the site, for example, power cuts or cuts in gas transmission, and the risk of failure to provide utilities, for example, steam, compressed air and so on, when raw energy supplies are available.

The former is a risk that is always present and will not be improved by outsourcing utility operation (although it may be improved by investment in additional supply points or backup generation, for example). In moving to outsourcing it would be unreasonable to penalise an outsourced energy service provider for a grid failure that is outside their control. On the other hand the risk of failure to provide utilities when there is provision of raw energy sources is a risk that can be transferred to an outsourced service provider. Any client company will attempt to put as much as possible of this risk onto the service provider, although this will be difficult or impossible when there are high costs of failing to produce, for example, in airport terminals when closing the terminal can cost the airport operator millions of euros in direct costs and in consequential losses. Appropriate liquidated damages should be negotiated. These should, wherever possible, reflect the true cost of failure but in many cases this cost could be too high for service providers to accept. The issue is better handled by recognising the need for redundancy, which leads to a requirement for additional capital, and flexible strategies such as including connections for temporary plant and putting in place contracts which will allow rapid delivery and installation of temporary plant such as boilers, air compressors and refrigeration compressors.

Supply-quality risk

The issue of the quality of energy usually only applies to power supplies, although there can also be issues with coal and oil if these are being used. Here the issue of power quality is discussed because of its increasing importance to modern manufacturing and building operation due to reliance on sophisticated monitoring and control equipment containing sensitive micro-electronics. This has dramatically increased the need for high-quality power.

Box 6.1 Power quality defined

The IEEE Standard Dictionary of Electrical and Electronic Terms defines power quality as 'the concept of powering and grounding sensitive electronic equipment in a manner that is suitable to the operation of that equipment'. Power quality may also be defined as 'the measure, analysis and improvement of bus voltage, usually a load bus voltage, to maintain that voltage to be a sinusoid at rated voltage and frequency'.

Source: Heydt, G. (1991)

Some equipment is particularly sensitive to fluctuations in power quality. In the past this has been mainly in high-tech facilities such as wafer fabrication plants or data centres but in recent years the spread of robotics and sophisticated monitoring and control systems in many buildings and industries has made this requirement far more common, even in apparently relatively 'low-tech' industries. The failure of monitoring and control equipment due to variations in power quality can lead to total production shutdowns. At the same time as the requirements for higher quality power are growing, the general quality of power supplied in the UK has declined (Briggs, 2003). Higher-quality power comes at a higher cost because of the need to invest in power conditioning equipment as well as monitoring equipment to control and measure power quality. The provision of high-quality power is a niche product area that power supply companies have generally ignored and represents an opportunity for ESCOs, particularly those linked to electricity supply companies.

Conversion-related risks

Efficiency risk

The efficiency risk is the risk of variations in the efficiency of converting raw energy into a useable utility; for example, gas into steam in a boiler, power into compressed air in a compressor or power into 'coolth' in a refrigeration plant. The conversion efficiency is driven by technical factors such the basic design efficiency of the equipment, how far it has drifted from that due to wear and tear, the age of the machine (in general efficiencies of new equipment have improved over time) and the quality of maintenance. In outsourcing contracts where a service provider takes over the ownership and/or operation of the plant this is a risk that reasonably should sit with the service provider who may in turn back some or all of it off to an equipment supplier or maintenance provider. The prospect of taking over this risk may drive the service provider to look at and invest in new equipment where the efficiency risk is reduced. It should also drive service providers to conduct detailed monitoring of the real world performance of the various types of equipment they are being asked to operate prior to contract in order to establish current performance. Post-contract, the service provider should continue monitoring on all equipment in order to ensure performance is maintained and develop a database of operating experience, which will be useful in assessing risks in any replacement/refurbishment decision as well as an aid to project development on other sites with similar equipment.

The service provider will attempt to contract at a lower conversion efficiency than the one they reasonably expect to achieve or have had guaranteed from an equipment supplier as this is one source of extra margin for the service provider, over and above that achieved at the base level of performance.

Demand risk

This is the risk of variations in the quantity of energy that the client uses at the point of demand. It is influenced by factors including:

- human behaviour, such as failure to switch equipment off when it is not required or using compressed air guns or high-pressure hoses to clean up the workspace;
- how the building or process plant is used, for example a supermarket adding more lights or LCD television displays for marketing reasons, changing opening hours, or opening windows in a heated or cooled building;
- changes in production volume;
- changes in product mix, for example a brewery producing a greater proportion of lager than ale in response to market changes (lager requires more refrigeration than ale in the brewing process, hence increasing power used per unit of production);
- changes in weather (average temperatures as measured by degree days) from the norm;
- variations in the efficiency of energy use/conversion in process plant.

Volume risk is not a risk that an energy service provider is usually able to take as the driving factors are almost completely outside their control. The only way that they can take this volume risk is if there is an agreed monitoring and targeting system that can be used to normalise energy consumption according to these exogenous factors. This is difficult even in a highly monitored situation and particularly difficult in large portfolios of buildings such as supermarkets.

Correlation risk

For any individual plant and utility there will be a correlation between utility demand and the level of activity or production. In outsourcing situations involving the sale of utilities, for example steam, compressed air, refrigeration and so on, the supplier will price each utility on the basis of an assumed or measured correlation. Therefore there will be a correlation risk for each utility, that is, the risk that the assumed correlation is incorrect. The degree of accuracy with which the correlation is known will depend on several factors including the extent and accuracy of past monitoring data, if it exists at all, which in the case of utilities such as compressed air and refrigeration is unlikely. An energy service company may determine correlations by detailed measurement and verification (M&V) during a due diligence phase. The accuracy of this will depend on the extent that variations in activity, or production, occur during the M&V phase as well as the accuracy of metering. Another factor determining the correlation risk will be the degree of correlation (R^2) as correlations with a low R^2 will unreliable for forecasting future demands. A further degree of uncertainty is the extent to which future production levels will fall within the range of production covered by the correlation.

Using an incorrect correlation can cause either under- or over-pricing leading to the service provider making less or more profit than it assumed when modelling the situation.

Operations- and maintenance-related risks

Capital replacement risk

Much energy and utility infrastructure is antiquated and has suffered from under-investment over the last 50 years with available capital naturally being directed to process and product improvements and innovations rather than utility systems. The risks are related to the age and condition of equipment, as well as its maintenance history, and it is unreasonable to expect a contractor to take on a large asset replacement/repair risk on aged plant and machinery. Some risks can be insured (at a price) but not usually this one. Options for sharing this risk should be considered. One option for reducing the risk is investment in new infrastructure, which has a price but will often lead to lower total operating costs.

Operations and maintenance risk

Operations and maintenance (O&M) risk can be divided into planned maintenance risk and unplanned maintenance risk. Planned maintenance risk (the risk that planned maintenance is insufficient to maintain the required operations) should be taken entirely by the service provider if, as would often be the case, O&M is in their scope. The unplanned maintenance risk (that is, the risk that a piece of plant fails unexpectedly, for example, if a boiler cracks a tube even though it passed inspections and has been properly maintained), is another more complex issue. Service providers will be reluctant to take this risk unless they specify and install new equipment or have a large portfolio of operating experience from which to assess the risk properly.

What is the extent of the risk, what equipment is to be covered, and what are the boundaries, are the key questions to answer at an early stage. It is important to be very explicit in defining exactly which items of plant and equipment are covered; areas that are sometimes neglected include distribution networks, underground assets such as pipes and tanks, high-voltage and low-voltage (HV/LV) distribution networks and building structures. If client companies attempt to push the risks of unplanned maintenance risk onto service providers there will be a price premium – a price made up of the projected cost of failure over the lifetime of the contract plus a risk premium to cover the uncertainty. Simplistic thinking along the lines of 'get the service provider to take all the risks' will lead to an increased total cost to the client as service providers seek to build in additional replacement capital. A balanced risk-sharing approach could lead to lower overall costs over the contract lifetime. If companies do go down this route service providers will have to develop more sophisticated asset management models and databases of performance of different asset types across their portfolio, as well as standardisation. Indeed, a service provider with a large portfolio of assets under management will be better equipped to take on some of the unplanned maintenance risks as the costs will be spread over a greater value of assets and revenue.

Project -elated risks

Investment in energy efficiency projects is usually considered 'low risk' but every project that involves installing or building something, whether it be relatively simple lighting controls or a large centralised energy centre with boilers and refrigeration equipment, carries project related risks, namely:

- the risk of not completing the project on time;
- the risk of not completing the project on budget;
- the risk of the project not performing as per the design.

All three types of risk apply to energy efficiency or utility infrastructure projects and outsourcing may represent an opportunity to pass these risks onto, or share them with, a service provider. The first two are construction risks and the third is performance risk. As such they are risks that can legitimately be taken by a service provider.

An additional risk that is sometimes not considered in energy and utility systems is technology risk, the risk that the technology is not sufficiently developed to perform to requirements. This is particularly relevant where organisations are trying to advance the state of the art by investing in emerging technologies before they are fully commercialised. Like all risks, taking it is acceptable if it is done consciously and with full acceptance of the possible outcomes. Most client organisations, who by definition are not in the energy business, are not well qualified to undertake energy technology development risks. Anyone contemplating using emerging technologies, for example fuel cells, would be advised to gain additional support and guarantees from equipment vendors or ESCOs offering such technologies. Being a test bed for technology development can be an expensive mistake.

The degree to which these risks are shared or transferred will depend on the nature of the contract and potentially affect the price that a service provider charges. A build, own and operate contract, for instance, will pass these risks onto the service provider and the provider will build a certain level of risk into their price.

Environmental and regulatory-related risks

There are several environmental-related risks around the use of energy and utilities. Again there will be local differences in regulations that affect these risks, but in essence all markets will have similar types of risks.

Climate Change Levy (CCL) risk

This issue is specific to the UK Climate Change Levy Negotiated Agreement scheme and is the risk of the company exceeding its CCL energy consumption targets and losing the Climate Change Agreement rebate (80 per cent of the CCL charge). For many industrial companies operating under a CCL Negotiated Agreement, the value of the rebate can be considerable.

EU Emissions Trading Scheme (EU ETS) risk

For companies operating within the EU ETS there is the risk that the company will breach its CO_2 emissions target and have to buy credits in from the market place or pay a fine of €40/tonne of carbon dioxide. The EU ETS also, of course, offers the prospect of selling excess CO_2 credits for those companies in credit and this is a potential opportunity or upside.

For companies operating outside the UK and the EU ETS there may well be other risks associated with emissions brought about by local carbon reduction laws and programmes. As we move into a world in which there are increasing constraints on carbon emission, whether they be by rationing or trading, there will increasingly be some kind of carbon emissions risk.

To fully understand this risk and how it changes it is necessary to monitor energy consumption, and hence emissions, and compare performance to the allocated level of emissions, on a regular and frequent basis. This may require upgrades to monitoring and targeting systems and an increased interest in energy performance from senior management.

Integrated Pollution Prevention and Control (IPPC) risk

As the IPPC regulations include energy-related items, and prescribe the use of best available technology (BAT), there is a risk of losing Environment Agency approval to operate if energy performance is poor or if the organisation is not using BAT or an acceptable alternative.

Brand and image risks

Although hard to quantify, risks around the brand and image of a company are large. Brands, and hence shareholder value, can be significantly damaged very quickly by environmental and product quality problems. In the future it is possible that 'being seen to be green' is likely to be an essential part of any brand image and any action or event that creates dissonance in the consumer's mind, suggesting that the brand is not as green as is being promoted, will have significant effects on sales. Being seen to be green is likely to be an increasingly important element of a brand as there is increasing awareness and pressure on issues such as climate change and oil peaking.

Examples of utility-related problems that could impact significantly on brand value include:

- the discovery of utility-related substances in products, for example, glycol from a cooling system in beer, leading to a general recall of product;
- a general release of a pollutant, for example, ammonia refrigerant

over the surrounding area;

- exceeding effluent consents and dumping waste into a river;
- injury or death caused due to a malfunction of a utility system or energy saving device which has been installed inappropriately.

Climate change risk

This is the risk that we plan based on one level of climate change occurring but actually what happens is different to the plan. This is a risk that is not generally considered at the level of individual organisations, but if climate change does accelerate, specific risks from climate change, for example the increased risk of flooding, need to be considered. There is also a larger business risk that entire business models may need to change in a carbon constrained world.

Risk transfer

An initial reaction to looking at this long list of risks could be, 'we will transfer them to a service provider therefore we don't need to know about them or understand them'. It could be said that utility outsourcing is a black box and the client should not care or know what is going on in that black box as long as they are receiving the required utilities to the required service levels. In the modern world, however, this approach is no longer acceptable. Health, safety and environment regulations and modern corporate governance procedures all require that the company exhibit a 'duty of care' even if all utilities are outsourced to a service provider. Therefore any outsourcing exercise is likely to be at best 'a transparent black box'.

The identified risks should be divided into:

- the risks that can be avoided;
- the risks that can be retained;
- the risks that can be reduced;
- the risks that can be transferred to a provider.

This risk analysis will feed into the desired deal scope and structure and hence the contractual form and content.

As well as the risks associated directly with energy, the risks of outsourcing and the risks of not outsourcing also need to be considered.

There are the risks of:

- continuing as you are;
- outsourcing and
  - not achieving performance at the desired level;
  - achieving no improvement in performance;
  - 'achieving' a reduction in performance.

The output of the risk analysis should be a risk allocation table showing the current situation and a desired situation after outsourcing.

THE QUESTION OF SCOPE – WHAT TO OUTSOURCE?

The scope of outsourcing can cover some or all of the energy management activities set out in Chapter 4. Table 6.1 sets out the various forms of energy management outsourcing and the activities which are usually included in the scope.

In deciding on the scope of the proposed outsourcing exercise, it is necessary to consider the analysis of core competencies, availability of internal resources and the desired risk profile.

To date most outsourcing in the energy arena has been selective with only specific activities within the energy management process being outsourced, most notably;

- monitoring and targeting through M&T Bureaux;
- operations and maintenance through O&M contractors;
- project identification and development through the use of consultants;
- construction of major utility infrastructure such as new boilers or combined heat and power (CHP) plant through contract energy management (CEM) or build, own and operate (BOO) contracts.

There are four 'value levers' in energy systems:

- the price of commodities (power and gas);
- the volume of commodities used (demand);
- operations and maintenance costs;
- capital expenditure.

In the past much of the value creation through activities such as CEM and BOO has come from transferring and reducing manpower and accessing greater buying power for O&M to reduce O&M costs, but not particularly from reducing energy demand. As most organisations have exploited all the potential to reduce O&M costs of their existing utility systems, the possibilities of achieving greater savings through reducing O&M costs are very small. CEM

Table 6.1 Typical scope of various energy management outsourcing forms

Outsourcing form	Activities included in scope
Strategic energy consulting including: • Benchmarking • Opportunity assessment • Policy assistance • Design programme	Recognise the opportunity Assess the risks Decide on strategic response Set policy Set high-level targets Establish responsibilities Allocate resources Ensure M&T is established Monitor results
M&T bureau	Establish and run M&T system
Consulting	Provision of M&T information Motivation and awareness training Identify projects Develop projects Manage implementation of projects
Energy procurement consultancy	Tendering energy supply, market advice Invoice validation
Build own and operate	Develop projects Implement projects Operation and maintenance Finance projects
Utility system operation and maintenance	Operation and maintenance of utility systems
Contract energy management	Operation and maintenance Provide M&T information Identify and develop projects Finance projects (usually supply-side for example boilers and CHP)
Energy savings performance contracting	Identify, develop and implement projects Finance projects or source finance
Utility outsourcing	Identify, develop and implement projects Finance projects Operation and maintenance

and BOO projects, where they did involve capital spend, have focused mainly on energy supply technologies, for example, new boilers and CHP plant rather than reducing energy demand.

Given the increased energy prices and carbon agenda, the future thrust of energy outsourcing must be in reducing energy consumption and carbon emissions through improving energy management and implementation of projects to save energy and replace out of date, inefficient utility infrastructure. This requires a more holistic approach to the total energy problem, allowing service providers to 'cross the meter' and become involved in end-use activities as well as supply activities such as running boiler houses of CHP plant. By doing this, organisations should be able to achieve greater value from outsourcing of the entire end-to-end energy management process, rather than continuing to select individual components of the overall process.

CONSIDER THE VARIOUS OPTIONS FOR PRICING MODELS

For any outsourced deal there are a number of approaches to pricing including:

- fixed price
- cost plus
- management fee
- hourly fee
- pay for performance
- gain sharing.

A fixed price transfers financial risk to the provider. It is useful where the scope is well understood and the volume is predictable, for example, M&T services where the volume of work is driven by the number of monitoring points or invoices received. Using it in any other situation will lead to either a high price incorporating every risk or a low price covering the minimum specification and a lot of change orders leading to extra costs. It is unlikely to be applicable to any complex outsourcing deal.

A cost-plus approach could be attractive with the proviso that costs are transparent to the customer and there are some other incentives to continuously reduce costs, for example, shared savings or gain share. Many of the early, first generation CEM deals in the UK were basically cost-plus without any incentives to reduce costs. There was little or no risk transfer and the failure to generate savings moving forward, coupled with the lack of cost transparency, has eventually led to widespread customer dissatisfaction with this approach.

A management fee approach can be applied to energy outsourcing. Costs that are outside the service provider's control, for example commodity costs, can be passed through to the client and a management fee charged on top of these. Again, unless there are other incentives in the contract such as gain share, this approach is not likely to be attractive. With a fixed management fee the supplier is taking the risk that they can deliver the scope within the budget.

An hourly or daily fee, which is the standard approach for charging for consultancy.

Payment for performance is at the heart of energy performance contracting (the most common form of energy outsourcing in many parts of the world, particularly the USA, but not widely used in the UK).

Gain share can work in energy service provision. One UK service provider operates purely on gain share on a simple formula of taking 50 per cent of the savings for two years instead of charging a consultancy fee. There are still, however, issues of measurement, coping with variations in energy price and variation on other factors that affect energy use such as production level, production mix etc. (See Box 4.4 on measurement and verification.)

Which is the best pricing mechanism will depend on the scope of the outsourcing exercise and the specific objectives of the customer (as well as the aspirations and intentions of the supplier). For a fully comprehensive service covering multiple aspects of energy management, a hybrid pricing system is likely to be appropriate, with different elements of the project priced in different ways.

The concept of large utility outsourcing projects in which long-term energy supply deals and energy demand-side investments are 'bundled' together allows them to be treated as structured finance deals. As with any structured finance deal a degree of financial engineering can be applied to pricing a utility project. For instance, fees can be spread over a longer period or front loaded to influence the cost profile that the customer sees. There may be good reasons to include some skewing of these costs in this way. Likewise, there may be good reasons why suppliers use these techniques to skew prices, such as demonstrating a lower first-year price. The project team need to be aware of these techniques.

CRITICAL EXAMINATION OF IN-HOUSE OPTION

Before finally deciding to outsource any activity, the options for keeping the services in house need to be carefully considered. Management need to critically examine whether the existing processes, programmes and suppliers can be re-engineered to achieve a better result with less effort and turmoil than outsourcing. The choice may not be 'outsource or carry on as normal'; more likely it will be 'outsource, re-engineer in house, or carry on as normal'. Carrying on as normal is only likely if the organisation is a paragon of virtue in the energy arena. Radical alternatives such as a joint venture with an energy service company, or a management buy-out of the utilities operation in conjunction with a service provider may be considered at this stage.

CONCLUSION OF THE DECISION PHASE

The completion of the decision phase leads to a proposal to outsource some or all of the energy management process, or to keep it in house. As this is an important strategic decision it would normally need to be ratified at board level and so the deliverable from the study normally will be a board presentation and paper. Assuming a positive decision in favour of outsourcing from the board, there is then a need to set up a small project implementation team to manage the process moving forward. As with the first stage, technical, financial and commercial people are needed in the team (along with a high-level sponsor), and it will help if there is some continuity between the evaluation team and the implementation team. The first task is to draw up an overall timeline. It is important at this stage not to underestimate either the time taken to undertake a full outsourcing exercise or the amount of management time it takes. An ideal is to second people to the project team and free them up from their normal responsibilities, but of course this is not always possible. As with the initial assessment, the services of a suitable consultant to guide the process could be useful at this stage.

PHASE TWO – THE PROCUREMENT PROCESS

Having decided to outsource, the next stage is to select a provider. This should be driven by the project team with the assistance of procurement specialists.

SELECTING PROVIDERS

The selection of service providers is divided into the following steps:

- identification of service providers;
- pre-qualification through issuing a request for qualifications (RFQ);
- issuing a request for proposals (RFP) or tender document;
- evaluation of responses to RFP;
- selection of preferred partner.

Identification of potential service providers means creating a long list of possible suppliers and then carrying out an initial assessment through an RFQ to produce a shortlist. Information on possible suppliers can be found from:

- industry trade associations, for example in the UK the Energy Systems Trade Association or in the USA the National Association of Energy Service Companies;
- other companies that have outsourced energy functions;
- utility companies;
- government energy efficiency agencies;
- general research using literature or web searches.

It is likely that some organisations may have been identified during the evaluation stage of the process and indeed their input may have been sought to clarify options and available models.

Pre-qualification should include assessment of the following characteristics:

- ability to deliver services similar to those required;
- provider strengths that match customer needs, for example sector experience;
- project performance;
- market reputation;
- customer satisfaction;
- strong financials;

- management capabilities;
- shared approach to problem solving;
- commitment to continuous improvement;
- transition experience;
- commitment of specific resources;
- trust/security/confidentiality;
- positive attitude;
- good cultural fit;
- flexibility;
- cost consciousness;
- willingness to share knowledge;
- clear vision of their marketplace.

Not all of these can easily be assessed by questionnaires. To properly assess them may require:

- inviting suppliers in to make presentations
- meeting suppliers at their offices
- visiting some of their operational sites
- talking to some of their customers.

It is important, particularly in assessing 'softer' issues such as cultural fit and good chemistry, to make a specific effort to meet a range of personnel from the service providers, not just their sales team.

Having developed a shortlist of service providers, a RFP or tender needs to be issued to the pre-qualified companies.

The RFP should:

- be clear;
- contain sufficient accurate information;
- give a reasonable time to respond;
- give access to decision makers.

In the document it is important to be clear about:

- the reasons behind the decision to outsource;
- the scope of the work;
- the required service provider qualifications;
- the required performance standards;
- the preferred pricing approach;
- the availability of decision makers (contact points);
- how questions will be handled (distributed to all bidders);
- any special terms and conditions that will apply.

There should be a clear focus on the issues and drivers that are important to the customer. It is important not to be too prescriptive in describing the solution that is desired as this may restrict suppliers from proposing the optimum solutions and impede innovation. Requests for proposals should be focused on outputs rather than inputs and exact methods.

The RFP should also ask the service providers:

- for confirmation that they can deliver the services as specified;
- to specify what, if anything, they cannot deliver;
- for any innovative ideas that could add value.

It is important to be clear about the approach to pricing and to cover:

- what is to be included;
- what is to be excluded;
- what format pricing should be presented in to allow as straight forward comparison as possible (including a pro forma);
- whether there are any start-up costs or whether there is any skewing of prices, for example, averaging of prices over the term.

It is also important to be clear about the methods of communications to be used during the procurement process. Basics such as:

- to whom communications should be addressed;
- the exact form of response required and any restrictions on identifying submissions;

- the fact that all questions and answers will go to all bidders;
- the exact deadline for submissions;

all need to be mentioned in the document. It should be stressed that the bidder should only talk to the designated contact or contacts in the client's project team.

EVALUATION

Despite all attempts to standardise the responses through standard pricing tables and other mechanisms, it is almost inevitable in any complex utility outsourcing project that there will be a variety in approaches to pricing. Variety in bids can be beneficial, but all proposals will need to be restated on common terms and assessed on a similar basis despite the difficulties of doing so.

In assessing the different bids it is necessary to:

- evaluate qualifications objectively;
- assess the confidence level that you have on the performance commitments made by the vendors;
- compare prices taking into account any:
  - different cost structures;
  - different risks;
  - different understanding of scope;
  - different assumptions;
  - different technological approach.

Remember, at this point in the process the price is not the primary issue. Price is negotiable and it is important to properly evaluate all aspects of the bids.

One of the key aspects that needs to be strongly reviewed is the provider's relationship manager or account manager. Special care should be given to reviewing and checking the qualifications, experience, approach and cultural fit of the person proposed for this position.

Once the evaluation has been completed each supplier should be invited to make presentations to the project team and other relevant stakeholders. The suppliers should be questioned with structured questions, focusing on open-ended 'what if' type questions. This is an important opportunity to test the 'chemistry' of the supplier's team.

Ideally the service provider will be fielding the team who will be doing the work, or at least senior members of that team, and questions should be addressed to those people rather than any senior staff or sales people present at the meeting. This requirement should be specified in the invitation.

Most outsourcing proposals can be summarised in a 30- to 60-minute presentation. Typically this would be followed by one to two hours of questions.

THE PROCESS FOLLOWING EVALUATION

The process moving forward from the evaluation stage is as follows:

- Confirm what due diligence is needed by both sides (how long this will take, what exactly is being proposed, what technical and human resources will be needed, what are the deliverables).
- Select the primary candidate (preferred bidder).
- Prepare a term sheet or agreement in principle.
- Make a public announcement.
- Prepare for the negotiation process.

DUE DILIGENCE

On any major outsourcing project, particularly those covering O&M and/or investment, there is a need to conduct due diligence. The purpose of due diligence is to confirm the validity of the vendor's assumptions in developing a model and pricing the project. In a complex energy outsourcing project, these assumptions are likely to include factors such as correlations between energy use and production for specific pieces of equipment which will require some detailed measurements to be taken over an extended time. The length of due diligence, the level of measurement needed and the costs, will vary between projects but a time of four to eight weeks, and a cost of £10 000 to £100 000 (depending on the amount of measurement equipment and duration needed) would be typical. The requirement for a period of due diligence should be included in the RFP. Normally the due diligence will be borne by the service provider. There also needs to be agreement on how the cost of due diligence is covered if the customer decides not to outsource. This agreement may be based on the savings confirmed by due diligence being similar to those identified in the proposal such that if confirmed savings are less than 85 per cent of those in the proposal the customer has the option not to proceed and supplier pays the cost of due diligence.

It is in both parties' interest that this due diligence is carried out well, as the success of any investment will depend upon it. As described in Chapter 4, traditional, simple 'energy audits' are insufficiently accurate to base investment decisions on. This requires due diligence of a standard similar to that of an investment grade audit. Due diligence is also an opportunity to refine planning for capital projects and to build working relationships between the customer's and service provider's teams.

NEGOTIATIONS

Before beginning the negotiation process it is important to be well prepared with:

- a list of all the relevant issues identified during the evaluation, ranking them in order of importance;
- an agreed strategy for the negotiation process;
- established ground rules for the negotiation process.

This is not intended to be a primer on negotiation, as there are plenty of texts dealing with this subject, but rather some points based on experience from negotiating complex energy services contracts.

Traditional, positional negotiating may work well for situations where the two parties never see each other again, but in a complex energy outsourcing project the two parties must work together going forward for many years. Fisher and Urey (2003) set out the idea of 'principled negotiation' and set out the four basic points of principled negotiation as:

- separate people from the problem;
- focus on interests rather than positions;
- generate a variety of possibilities before deciding what to do;
- insist that the result be based on some objective standard.

The overall process should be one of analysis, planning and discussion. Getting into discussions without the necessary analysis and planning will cause problems and delays, which will frustrate both sides. The best way to handle the logistics of negotiation is to set aside concentrated blocks of time (say two or three days at a time) for discussions, allowing time for analysis and planning in between the negotiation sessions. Basic meeting courtesies should always apply, for instance staring meetings on time. Having meetings off site, away from the day-to-day pressures of the office, will help in this regard.

A draft term sheet listing the major points to be included in the contract should be put together. It should cover the major items of scope at a high level rather than attempt to capture every detail. Once it has been prepared and checked it should be sent to the vendor with a request for a formal response on each point. Following the response, negotiations on the term sheet should commence. A typical term sheet would include the following:

- scope of services
- contract term
- performance standards
- pricing
- factors of production
- management and control
- transition provisions
- billing and payment terms
- termination issues
- confidentiality
- warranties and indemnities
- limits of liability
- use of sub-contractors
- third-party licences.

Usually a term sheet will only be negotiated with the selected provider but this will reduce the purchaser's commercial advantage. If two providers have been evaluated as being close in capabilities and skills, then it is normal to hold one on stand-by until negotiation of the term sheet has been completed to retain commercial advantage. Once a satisfactory term sheet is agreed the decision to transact with the vendor should be confirmed and publicly announced.

The draft term sheet should feed into an issues 'punch list', which summarises the issues, the status of agreement, the details of any debate and any agreed actions along with who has responsibility for them. Colour coding these helps so as progress is made it can be recognised by turning the 'traffic lights' green. Circulating this to all members of both negotiating teams helps to keep people focused.

After the major items are agreed in the term sheet there will always be a range of issues, ranging in impact and importance. Fifty per cent of the issues are likely to be 'throw aways', meaning they have no significant impact; 25 per cent will have minor impact and 25 per cent require real negotiation. The focus should be on the latter.

Negotiations will be helped if at early stage agreement can be reached on an open-books approach and to use a transparent, commonly agreed pricing model which will allow both sides to focus on the cost structures and effects of any assumptions being changed. Clearly this requirement should be stated in the tender documents so that it is not a surprise to the preferred vendor.

At all times in the negotiation process both sides should remember the principles of fairness (Greaver, 1999), which are:

- the customer deserves the best service;
- the provider deserves to get a fair price;
- the customer deserves a fair market price for any assets being sold;
- mistakes will occur on both sides;
- the aim should be for a win–win contract;
- there should be a perception of fairness on both sides.

Negotiations for complex energy services projects will take considerable time, typically three to six months to negotiate. As the outsourcing project is long term it is in both parties' interests to negotiate a fair agreement. Re-negotiating, either with the selected provider, or another if the relationship breaks down, is expensive and undesirable for both parties.

CONTRACTING

Contracts for outsourced energy services can be complex but in general can be divided into eight sections:

- scope of services
- service level agreements
- pricing schedules
- factors of production
- management and control
- transition provisions
- terms and conditions
- termination provisions.

For complex transactions there will be several appendices, covering for example:

- the list of specific assets that are being transferred to the service provider
- detailed service level agreements (SLAs) for each service.

The ideal situation, of course, is one in which the two parties put the contract away once it is signed and get on with the job in hand. This, however, is only possible if the agreement is well drafted and comprehensive. Many original outsourcing, contract energy management or energy performance contracts have foundered because the contracts were badly drafted or simply assumed that some issues could be left to future agreement. Although it is essential to allow for future agreement between the parties on issues that cannot be clearly defined at the time of the contract, and for what happens if agreement is not reached, critical issues should be defined, agreed and incorporated into the agreements.

SCOPE OF SERVICES

The scope of services should have been defined through the RFP stage and negotiation of the term sheet. It is useful however to start the process by testing the mutual understanding of the scope of service by developing 'what if' questions around the edge of the scope of services. Asking these questions and developing a common understanding is a crucial stage in capturing the full scope within the contract documentation. Differences in expectation and understanding on issues of scope will occur between the parties and will need to be aired and then agreements negotiated. The resulting list of services, functions and activities should be comprehensive but avoid falling into the 'task' level details trap. The use of precise language brings comfort to both sides in setting out exactly what is required, but can be limiting in a situation in which the business environment is ever-changing.

DECISION RIGHTS

In principle the client should be less concerned with the processes and methods than the outcome but the realities of duty of care mean that the client will want to retain decision rights over certain aspects, typically:

- the provider's account manager and key direct reports (the client needs to have confidence in their abilities and trust them);
- any voluntary movement of personnel to other contracts;
- any significant changes in technology;
- new project specifications (the client needs to be confident in the project technology, reliability, safety and environmental impact);
- any changes in outputs or deliverables, for example report formats and frequency;
- any changes in processes, for example anything that could affect internal processes or impact on regulators or the public.

Major issues of scope are considered below.

THE SUPPLY OF UTILITIES

The provision and supply of any utilities which are included in the transaction need to be explicitly described.

It is normal in utility provision contracts to define SLAs for each utility, which cover quantities, qualities and acceptable downtimes. The question of what happens if demand for utilities exceed the available supply needs to be considered and answered. This can be addressed by specifying a recovery period after the maximum capacity has been exceeded, during which the obligation to supply the utility is suspended.

In addition, events that trigger suspension of the SLA (a 'suspension event') need to be defined. As well as demand exceeding capacity, other acceptable suspension events would include, for example, grid power failures.

For each utility the method of measurement and metering need to be considered. An example of the kind of issue that could come up here is the metering of refrigeration. A heat meter could be installed, but it may be considered by both parties more economical to meter power into the refrigeration system and assume a standard coefficient of performance (CoP). This is because heat meters, particularly for large refrigeration installations, tend to be expensive to install and unreliable.

The ownership (and responsibility for maintenance) of meters needs to be defined, particularly where there is an asset transfer of utility plant and distribution systems. In addition calibration of meters needs to be considered. There should be an agreed schedule of recalibration for main utility meters, particularly if they are fiscal meters (being used for billing purposes). The contract may also give both parties the right to challenge the current calibration if a major error is suspected. For instance, if a party challenges the meter calibration, the challenging party should pay for the meter testing and recalibration. If the 'allegation' is found to be correct, the other party should reimburse the challenger's costs of testing and recalibration.

The physical positioning of meters is an important issue, as it tends to drive certain behaviours. In traditional energy services contracts where utilities are sold, for example a BOO boiler house selling steam, the utilities were always metered at the outlet from the boiler house. The service provider then is only responsible for provision of steam to that point and ownership of the steam transfers to the client at that point. Assuming the service provider is being paid by the tonne of steam delivered, with no consideration of the input energy used, this drives the service provider to maximise energy efficiency within the boiler house, that is the conversion efficiency. It does not, however, incentivise the service provider in any way to consider the situation 'over the fence', such as the efficiency and effectiveness of steam distribution and use. In fact, it is in the service provider's interest for the client to be inefficient and ineffective in its use of steam. The more steam leaks there are in the distribution mains, and the more inefficient any heat exchangers or heater batteries are, the more steam is sold and the more revenue and profit the service provider makes.

To correct this disincentive, it is necessary to move the point of metering to the point of use, that is just before the end use. This presents practical problems in terms of there usually being multiple end uses, for steam in a brewery for instance, there could be tens of end-use points. Metering each is likely to be expensive and create a meter maintenance problem. It would, however, have the benefit of incentivising the service provider to at least consider the efficiency and effectiveness of the steam distribution system, and accompanying condensate return system. With this kind of arrangement, maintenance of steam traps, repair of leaks and installation of high-efficiency orifice steam traps are likely to be of interest to the service provider. It should be noted, however, that it still does not incentivise the service provider to consider the effectiveness and efficiency of the actual end use process. That requires additional mechanisms in the contract.

Linked to the physical location of meters is the issue of transfer of title of the utility being supplied, meaning when does the steam or coolth or compressed air being sold to the client actually pass into their ownership. As well as fiscal implications this could also have environmental or health and safety implications.

For each utility there will need to be a set of billing assumptions (see Figure 6.3). These are concerned with practical engineering matters such as the physical location of meters as well as the variables that drive pricing. For instance, if a utility such as treated water depends on an input commodity, for example mains water whose price cannot be controlled by the service provider, the billing assumption for treated water will include a formula linking the price of treated water to the price of the input water. The price of input water is a billing assumption.

COMPLIANCE WITH STANDARDS AND REGULATIONS

The client will need – and indeed has a duty of care – to ensure that a utility service provider follows all appropriate standards and regulations, for example the Legionella L8 standards. An outbreak of Legionella (Legionnaires'

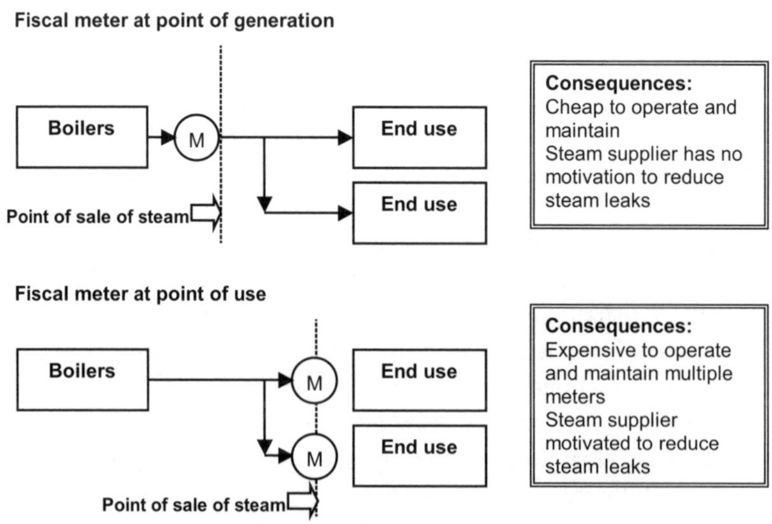

Figure 6.3 Fiscal metering points

disease) would cause a major problem in terms of publicity and brand image for both parties, as well as direct legal problems from the regulatory bodies and possibly those individuals affected. Outsourcing does not shield the customer from responsibility in areas such as Legionella and effluent consents. The customer has a duty of care to ensure that the service provider has appropriate processes in place and that these are being adhered to: appropriate safeguards need to be built into the contract. This kind of monitoring can be assisted by openness of IT systems and web-based monitoring systems which allow other users (whether in the service provider or the customer) to view performance. Exercising duty of care is also assisted through using periodic, independent audits, which should be allowed for in the contract.

SCHEDULING OF PLANNED MAINTENANCE

The provision of planned maintenance is usually a central part of any utility outsourcing contract. There is no point having a service provider causing disruption by scheduling planned maintenance during times of peak production. There is clearly a need to ensure that co-ordination and mechanisms of co-ordination, for example attendance at engineering meetings and requiring client approval of any planned maintenance schedule, are defined and included in the contract to drive the service provider to the required behaviour.

UNPLANNED MAINTENANCE

The risk of unplanned maintenance is a major one in any utility operation. Who takes this risk, the client or the service provider – or indeed whether there is some form of risk sharing – will depend on the type and age of the equipment, the O&M experience, and the aspirations of both parties. It is vital, however, to clearly define what constitutes unplanned maintenance and who is taking the risk in all energy service contracts.

Equipment failure insurance may be available for newer equipment but is not generally available for older, existing equipment. Also, new equipment will come with some form of manufacturer's guarantee, usually one year at a minimum, and the service provider and/or customer can explore the costs of acquiring longer guarantees from the manufacturer (which may be tied into maintenance being undertaken by the supplier).

As long as the service provider is carrying out routine O&M correctly, unplanned maintenance is a 'no fault' event which is outside the control of the service provider and therefore it seems inappropriate to try to pass the risk of unplanned maintenance onto the service provider directly (unless it is for new equipment). Newer methods of risk sharing, such as jointly contributing to a 'risk fund' on an annual basis may be appropriate. In situations where the client is running all their own utility services they bear the risk of unplanned maintenance.

Rather than trying to pass unplanned maintenance risk onto a service provider, scarce management and engineering time would be better focused on improvement projects that will reduce the risks of failure of major pieces of plant. These may include increasing redundancy where possible (for example having two water or power sources), or adding in connections for rental equipment such as boilers and air compressors and putting in place contracts that ensure rapid access to such equipment. Such work should be based on a full risk assessment that identifies potential single point failures and the risks associated with them.

PROJECTS

Many utility outsourcing contracts will incorporate the identification, development and implementation of capital projects to improve energy efficiency, upgrade or replace utility system infrastructure, reduce O&M costs, increase (or decrease) utility system capacity, or increase availability and reliability. One of the attractions of outsourcing is to have these projects identified, developed and implemented by a specialist energy services company. Another major attraction of outsourcing is for the service provider to actually invest in these projects in return for a share of the benefit, however this may be structured.

One approach to outsourcing involves the service provider identifying key projects during the competition phase and indeed, the quality, quantity and potential benefits of the projects identified may be a major differentiator between service providers. The financing of these projects can be handled in a number of ways with either the client or the service provider accessing the capital. Repayment of the costs can either be through a separate capital charge or, in the situation where the utility provider is selling utilities, incorporated into the utility rates. This transfers the project completion risk and the project performance risk to the service provider who is then incentivised to ensure that the project is built in time or quicker, and that the project at least meets the performance expectation built into the utility rates.

There are four dimensions to consider in thinking about capital projects These are:

1. Who is to identify, develop and implement projects – the client, the service provider acting as a developer, the service provider acting purely as a consultant, both parties working together, or a third-party consultant?

2. Who is to finance the projects – the client, the service provider or a third-party bank or finance company? Either option will have different implications for asset ownership, which risks are to be taken by whom, and the level of net benefits to both parties.

3. How is the value stream from the projects divided between the parties? What division of benefits is appropriate and over what period?

4. When should projects be identified, developed and implemented? Should they be developed at the beginning of the contract and built into the contract or should they be developed after the start of the contract?

If investment in projects is part of the scope of services required, however they are being funded, there will need to be close co-operation between the service provider and the customer on the scope and programming of the projects as well as the technical and organisational interfaces. The client will need to practise a degree of oversight over the planning, design and implementation of capital projects (and indeed non-capital projects such as training and awareness campaigns). This is needed to satisfy the requirement to exercise duty of care for issues such as health and safety and environmental impact, as well as the practical need to integrate utilities with end-use processes or buildings. There is also a need to ensure that there is no disruption to production or use of the facility as connections and tie-ins are made; therefore these events need to be agreed and co-ordinated between the parties.

The investment phase of the outsourcing cycle will need enhanced and expanded resources from both sides, for instance additional engineering or project management staff, reporting to the programme manager and account manager respectively. A projects team with representatives on both sides, engineering or project managers, reporting to the programme manager and account manager, can be useful at this stage. The focus of both sides should be on getting the projects built on time, and on cost with minimal disruption.

SHARED SAVINGS PROJECTS

Conventional build, own and operate (BOO) and contract energy management (CEM) deals are based on the sale of heat and/or other utilities. This arrangement can actually incentivise the service provider to provide more utility rather than less, a perverse and inappropriate incentive from both the point of the view of the client and the need to reduce emissions of carbon dioxide. Some of the more innovative projects, for example RWE – Diageo Utility Alliance Agreements, include a shared savings mechanism which motivates the service provider to seek out and implement energy cost reduction measures which can either be in the utilities themselves, or 'across the fence' in the process end-use side. The contract needs to be structured in such a way that ensures that it is more profitable for the service provider to identify and invest in energy saving projects than to continue to provide the original volume of utility. This means that, as well as providing an acceptable rate of return to the service provider, the share of the benefits passing to the service provider has to outweigh the value lost through a reduction in utility sales.

SERVICE LEVEL AGREEMENTS (SLAS)

Clear service level agreements (SLAs) are essential to the success of any outsourced project. Often, internal organisations providing energy services are not subject to SLAs and the beginning of the outsourcing process will soon lead to the question, what service levels should we apply? Often the application of any SLAs can lead to improved performance. In contracts covering the supply of utilities SLAs will generally cover the three issues of quality, quantity and availability.

It is necessary to have a pragmatic approach to determining SLAs. In many cases existing SLAs are satisfactory and there is no point in seeking to improve upon these. In other cases there will be clear deficiencies in existing SLAs, which can only be corrected by investment. In this case temporary SLAs may need to be redefined and post-investment SLAs either defined within the contract or a mechanism to agree incorporated in the agreements.

For outsourcing projects involving the delivery of utilities the maximum supply capacities should be defined as well as parameters such as recovery time after an event where demand exceeds maximum supply capacity. For instance, it is possible to draw 26 tonnes of steam per hour from a boiler installation with a maximum supply capacity of 24 tonnes/hour but this will draw down water levels in the boiler, thus limiting output for a period of time after the event.

A mechanism for varying the SLAs by process of agreement should also be defined. Changes in circumstances, or indeed operating experience, may lead to the need to change the SLAs for one or several utilities. Clearly, however, some changes will have cost implications for example the need to invest and therefore cannot be imposed unilaterally by the client. An improvement in an SLA requiring investment will have an implication on the price of the services.

An infringement of an SLA needs to be defined as an event. Some infringements of SLAs may be deemed to be less serious than others. In this case the types of event, minor and major, need to be appropriately defined in the service levels. Minor events could be, for instance, changes in steam pressure that in themselves do not create a production problem whereas a major event would be a drop in steam pressure that leads to a production failure.

If minor and major events are defined there should be an agreed escalation procedure that takes care of the situation where a repetitive minor event may not cause any production problems but probably reflects a larger underlying problem, which the service provider is not getting to grips with. This situation could be covered by the concept of a subsisting minor event. Such a subsisting minor event may count as much as a major event in terms of liquidated damages (LDs) and contract key performance indicators (KPIs).

KEY PERFORMANCE INDICATORS (KPIS)

As well as operational SLAs there is a need to have contract KPIs. These may include overall measures of performance such as conversion efficiency, overall site energy efficiency and speed of response. The KPIs are more concerned with the overall management of the contract than the individual technical SLAs.

LIQUIDATED DAMAGES (LDS)

Liquidated damages (LDs) limit the liabilities of the supplier and therefore are likely to be required by both sides, by the supplier as a way of limiting liabilities and by the customer as a way of reducing the likelihood of negative behaviours. The values of LDs should be related to both the cost of damages and the total profit on the contract. A number of potential clients, when looking at utility outsourcing, have attempted to impose completely unreasonable LDs on service providers which, although they reflect the true costs, for example the cost of shutting down an airline's check-in desks or the cost of not being able to supply utilities to a major automotive company, are out of proportion to the total potential profits of the service provider. These are almost certain to be rejected by any serious service provider and any attempt to impose them will simply lead to a much higher price for the services. The aim should be to establish a regime in which damages increase for a pattern of repeated failures and are significant enough that the service provider will not want to incur them, but not so large that they add an unacceptable risk premium to the price of the project. Procurement departments still often seek to impose unlimited damages and consequential loss on utility service providers. Corporate governance guidelines in most large utility companies or energy service providers will prevent these being agreed to and so asking for them is a waste of both parties' time.

Companies that have examined the risk issues in some detail and pioneered risk sharing approaches such as BAA seek to ensure risks are minimised in other ways, for example design of systems with built-in redundancy and careful selection of service provider.

PRICING SCHEDULES

The method of pricing needs to be set out clearly. As described above there are alternative pricing mechanisms for utility outsourcing including:

1. fixed price, for example per unit
2. cost plus
3. management fee
4. hourly fee
5. pay for performance
6. gain sharing
7. value based
8. other.

The assumptions behind any pricing structure, as well as the impact of any change in these assumptions, need to be defined. These assumptions, for instance, may include the price of bought-in commodities such as industrial gases, power or gas, which are outside the control of the service provider. Clear formulae which show the impact of a change in any of these pricing assumptions need to be developed and set out, probably in an appendix.

Other external factors such as the impact of cost of living increases and the use of indices such as the retail price index (RPI) or consumer price index (CPI) need to be considered. If prices are to be linked to an index such as RPI it is important to be clear about whether it is the correct index – consumer price inflation can be different to industrial input price inflation, for example – and to which elements of price it refers to. Applying a blanket index to a utility price that includes many elements such as fuel, O&M, labour, and capital invested, for instance, will lead to excessive increases and additional profit to the service provider as the capital element is not affected by inflation.

FACTORS OF PRODUCTION

Facilities to be provided by the customer

In most outsourcing projects it is normal for the client to be obliged to provide some facilities and services to the service provider, including defined office space, telephone line(s), IT network access, a control room if appropriate, cleaning services and car parking. These facilities need to be defined along with any restrictions that are to be imposed on their use.

Factors of production to be transferred

Some utility outsourcing projects include the sale of existing assets to the service provider. The assets that are to be transferred need to be accurately and completely defined in the form of an asset register. The price of these assets (usually book value) need to be agreed, as well as whether any warranties or guarantees will also be transferred. Both parties need to ensure that they are satisfied that they understand any tax issues associated with asset transfer, and good tax advice is vital in this regard.

Provision of rights of access

The client needs to grant the service provider rights of access to their site. In many cases utility equipment will be distributed through the site and so this right should be extended to the entire site. In some cases the client may want to restrict access to particular areas where, for instance, there is confidential research and development activity. Any exceptions such as this need to be defined in the agreements. These rights of access will often, depending on the legal jurisdiction, be defined in a property lease. Any restrictions on access, or special requirements, such as security screening should be included as well.

Licence to use systems

If the utility assets, or any part of them, are not being transferred to the service provider but have to be used by the service provider, for instance to carry out O&M or simply to provide the utilities, then the client needs to grant the service provider a licence to use these systems. The terms of the licence need to be defined and agreed as well as the circumstances in which the licence can be or will be revoked for example termination of the contract.

Allocation of electricity and gas

In some outsourcing projects the service provider will be responsible for buying the electricity used in the production of the utilities, while the client remains responsible for the purchase of power for process use. This is helpful for driving behaviour on both sides to improve conversion efficiencies. Usually the layout of main incoming power meters and HV and LV distribution systems will not allow a separate power supply contract to be entered into, or for a simple allocation of meters. Therefore the allocation of electricity needs to be defined and set out in the contract. This could be in the form of a separate agreement or a schedule to the main agreement but a clear definition of meters, and responsibility for consumption and payment needs to be agreed. This may involve tripartite negotiations with the electricity provider.

Gas use in utilities is usually more easily defined and metered than power. For instance it is possible to install, if they are not already installed, separate gas meters on supply pipes to boilers (in a group or individually) and CHP installations.

Contracts for provision of other services and/or commodities

In order to provide the services required some contracts for factors of production will need to be novated to the service provider or new supply arrangements put in place. These may include items such as the supply of industrial gases, effluent consents and chemical treatment supplies and testing. Novation of these contracts or permits to the service provider will need to occur during the transition period between contract signature and the commencement of services.

MANAGEMENT AND CONTROL

Organisation

Many energy services contracts will run for extended periods of five, ten or fifteen years, some have even longer terms. The management of the relationship cannot be left to chance and relationship structures need to be defined in the contract.

The contact points within the respective organisations need to be defined as well as escalation procedures. Contact points should be at several levels covering operational issues, contractual issues and strategic issues. An example of appropriate contact points is shown in Table 6.2.

Table 6.2 Examples of client–service provider contacts

Contact Level	Client	Service Provider
Operational	Engineering Manager	On-site Utility Manager
Contract management	Contracts Manager (Programme Manager)	Account Manager
Strategic	Engineering or Commercial Director	Account Director

There should be a single point responsibility for the contract on both sides, with a programme manager from the client and an account manager from the service provider. To oversee the contract, however, there should be a client steering committee of defined membership from relevant parties including technical, production and finance. The steering committee is effectively the 'board of directors' for the on-going contract and relationship. The chair of the steering group should be the executive who the service provider's account manager reports to, for example, the programme manager. It would be prudent to include the steering group in the contract, as it is important for managing the on-going relationship over what is usually aiming to be a long-term relationship. It also helps to give continuity and stability as inevitably individual members leave or change job functions and rotate into other areas.

The frequency of meeting for the steering group will be higher during the busy transition phase and the early stages of the contract. Depending on the details, once a month during the first year and then reducing to once every six months or once a year is probably an appropriate meeting frequency.

Below the steering group level the specific, contractual meetings between the client and the service provider must be defined in the contract. These are the management meetings for the contract and would normally be between the client's programme manager and the service provider's account manager.

The items to be discussed at the contractual meetings should be defined in the contract and are likely to include:

- analysis of performance reports;
- discussion of performance issues;
- analysis of customer issues ;
- analysis of provider issues;

- resolution of problems;
- report on capital projects (if appropriate);
- report on continuous improvement projects;
- recommendation for changes;
- planning new initiatives;
- reviewing any significant changes;
- reviewing billing.

This management meeting should normally be held monthly with more frequent meetings during the transition. The agenda can be standardised and meeting dates, times and locations should be agreed six to twelve months in advance so that they become routine.

The objective of the meeting should be to jointly address problems and share in solving them, not to assign blame. Problems should be resolved at the lowest levels possible and delegated powers can be established with set limits for each level. In the event of disagreement that cannot be solved between the parties the issues need to go to mediation or arbitration. The methods for mediation and arbitration need to be set out in contract, possibly with timescales. Although in an ideal world both parties should never have to use these clauses they need to be familiar with them.

In most cases of energy outsourcing it is also necessary to include the service provider in other meetings such as regular engineering management meetings and production planning meetings, as there will be a need for interfaces between the energy services providers. These functions and utilities need to be closely integrated for optimum results. Other relationships, including human resources (HR) (for initial Transfer of Undertakings of Public Employees regulations [TUPE] activities and recruiting and site-wide energy training for instance) need to be established as well but are not likely to be so frequent.

Periodically (every six months) the customer and service provider should meet at senior management meetings, strategic management meetings or even board of directors meetings. These meetings should consist of sharing issues and not just 'beating up' on the supplier.

Performance reports, including those covering energy consumption, need to be disseminated in a standard, agreed format.

Reporting and meetings

The nature and frequency of reporting should be defined precisely. What is to be reported, to whom, how often and by what means, all need to be defined in the contract.

However much the client may want to discharge themselves of the responsibilities that accompany operating energy and utility plant, the requirements for duty of care mean that energy outsourcing deals have to be 'transparent' rather than a 'black box'. As well as operational reporting there needs to be a mechanism by which the service provider demonstrates to the client that they are carrying out statutory responsibilities in a suitable manner. This mechanism could be by regular review of written procedures or by a formal, independent audit once a year of relevant procedures and systems.

Change management

In any long-term outsourcing project it is inevitable that there will be changes in circumstances that require changes to the services offered. Changes in production levels, production mix or production technology will affect the level of demand for utilities and may indeed require additional investment in utilities equipment. Processes for change management should be considered. Some changes will be within the level of devolved responsibility of the account manager and the programme manager, whilst others, particularly if investment is required, will require re-negotiation which will take time and require dedicated resources from both sides.

TRANSITION ARRANGEMENTS

Transition period

The agreement of a suitable transition period, that is the period between contract signature and the commencement of services, and what activities have to be completed by both parties within that period, is important. Outsourcing of energy and utilities is too complex to have everything happen at once and there has to be an agreed transition plan to move from existing arrangements, whether in-house or through an existing outsourced contract, to the new arrangements. This is even more important where the transfer of staff is included as each party needs to ensure satisfactory completion of their statutory obligations in respect of employees which come about through the Transfer of Undertakings Protected Employment (TUPE) regulations. Having qualified and experienced TUPE advice is clearly important.

The activities that need to happen in the transition period will vary but the following are typical:

- the TUPE process with all necessary employee consultation (assuming employee transfer is part of the scope);
- operations manuals need to be written;
- HV switching procedures need to be agreed;
- crisis management plans need to be agreed;
- procedures for formal handover of assets and transfer of responsibility need to be agreed (assuming asset transfer is part of the scope);
- information on the process needs to be given to the rest of the site/ company;
- recruitment of staff;
- training of existing service provider staff;
- any necessary signage needs to be prepared;
- liaison with statutory bodies such as the Environment Agency to transfer effluent consent.

What is the ideal length of transition period? There are two conflicting forces at work; one is the tendency of the transition team to want as long as possible, and the other, usually more powerful force, is that negotiations drag on longer than planned, preventing starting work on the transition plan, while the 'go live' date is fixed, leading to truncating the transition period. The ideal length of the transition period also depends on the scope and complexity of the project. Thirty to ninety days is normal and this needs to be included in the overall project programme at the start of the process, otherwise the transition period will end up being compressed between the tendency for negotiations to drag on and the need, real or perceived, for an absolute 'go live' date.

The contract will usually include obligations on both parties during the transition period. These may include the following.

An obligation on the customer to continue operating and maintaining the utility assets that are to be transferred to the service provider, either by ownership or simply for O&M. The standards of O&M need to be defined here and appropriate safeguards put in place to prevent a running down of maintenance during the transition period. However tempting it may be, a short-term saving here for the client is likely to result in longer-term problems and costs.

If a transfer of personnel is involved, particularly under the TUPE regulations, there needs to be an obligation on the customer not to make any changes to the number of employees to be transferred or their terms and conditions of employment. This is important, as the service provider would have priced their services on the basis of a defined number of staff and defined terms and conditions.

There may be a requirement on the client to install monitoring devices such as meters and sensors prior to the services commencement date so that performance monitoring and invoicing can be up and running from the first day of the contract. If so it is important to agree that the service provider has the right to access the site, make the necessary installations and monitor the equipment. Likewise the client will want to impose obligations on the service provider whilst installing equipment such as the need to co-operate in timing installation work to minimise disruption.

Both parties should be obliged to put appropriate, agreed insurances into effect at or before the services commencement date and to prove that they have done so through contacts between insurance brokers or representatives.

There should be an obligation on the service provider to draft and gain the clients approval for an operations manual with the client. In the first instance this could be simply a 'rebadging' exercise of existing manuals to save time and ease the transition, but if so there should be a defined timeline for preparing and agreeing a revised manual. The frequency of updates for example annually or every six months, should also be agreed.

In a similar way there should be an obligation on the service provider to prepare and gain the client's approval for a crisis management plan. Given the essential nature of utility services such as high voltage electricity, refrigeration, steam, water and effluent, and their potential for site-wide and indeed local area, health and safety and environmental problems, it is vital that the crisis management plan is fully integrated with the client's existing plan. This is particularly important in sections covering accidents and environmental emissions as a joint approach to the crisis and the management of public relations can make all the difference to the effect that any incident has on public perceptions.

The culmination of the transition period is the commencement of services, which will usually occur at a defined time, for example 00.01 on a date defined in the contract. If all the transition activities have been completed successfully the actual transfer of responsibilities from client to service provider should proceed smoothly without any glitches.

Extended transition period

In order to deal with the practical issues associated with handover of assets, transfer of staff, and the novation of existing third-party supply contracts, the parties may want to define both a transition period, the period between contract signature and services commencement, and an extended transition period, a period starting after services commencement. This may run up to 90 days and allows the service provider to complete activities such as proposing and agreeing new operations manuals, and changing staffing patterns where allowed by TUPE.

The operations stage

This should be a time of stability and normality as the transition has been made successfully, new investments have been made and any relationship or administrative problems should have been worked out. The emphasis in this stage of the process is clearly on routine operations although inevitably, with the rate of change in modern organisations any period of stability may be short-lived. This makes it important to have agreed processes for change management.

Ideally the contract structure should also incentivise both parties to seek out and implement continuous improvement (CI) projects and these should be the main focus of activity apart from any occasional equipment failures or departures from SLAs. CI projects are very powerful, both as a way of reducing energy costs and a way of motivating staff.

TERMS AND CONDITIONS

As well as normal commercial terms and conditions covering issues such as payment terms this section will include other items. These include:

- insurances
- review period
- exclusivity.

Insurances

The types and scale of insurances to be taken out and maintained during the life of the contract by both parties should be defined in the agreement. These will typically include:

- public liability (mandatory)
- professional indemnity
- business interruption.

Review period

Clearly within a long-term utility outsourcing contract, many things can and will change including the client's and the supplier's staff, their organisational structure, the client's products or product mix, the client's location, the available utility technologies and last, but not least, energy sources and energy prices. It is prudent to have a strategic review built in to long-term contracts, for example every five years, which allows both sides to consider whether the arrangement is truly working for them. Too frequent a review period will tend to invoke uncertainty and tend to drive unwanted behaviours such as running down O&M or investment.

Exclusivity

For any major energy outsourcing project it is almost certain that both the client and the service provider will want exclusivity. If there are any exceptions to the exclusivity they need to be negotiated, agreed and incorporated into the contract. An example may be where there is an existing outsourced CHP on the site, or where the client wishes to continue to use an independent metering and invoice verification bureau.

TERMINATION PROVISIONS

Any contract should consider and define the possible causes and the consequences of termination. Termination can normally be of three types:

- situations where there has been a default on the part of one of the parties for example failure to meet SLAs consistently;
- situations where there has not been a default for example closure of a site or take-over of either the client or the service provider;
- expiry of the contract term.

The different types of termination and what could lead to them need to be defined. For instance, if a service provider fails to meet the service levels consistently over a certain period, or has more than a set number of major events in a year, these could be defined as service provider default termination events. If the client fails to pay its bills for three months this could be defined as a client default termination event. It is important to consider many scenarios including:

- either party going into administration or liquidation (however unlikely this may seem in retrospect, remember at the time the collapse of Enron was generally unimaginable in 2000 and indeed up until late 2001 for those involved with the company and their customers);
- a failure of both parties to agree a change to the contract;
- a failure to perform basic service level agreements;
- change of ownership of either the service provider or the client (this may affect parent company guarantees);
- reaching the end of the contract.

For each of the types of termination, including the simplest and ideal one of reaching the end of the contract term, it is necessary to consider the implications in terms of the financial obligations on each party as well as the actual logistical problems that termination will cause. If assets have been purchased or investments in projects made by the service provider there will be a termination fee designed to recover that capital. Payment terms for this amount must be defined. If operating staff are involved, the intended path of transfer, back to the client or onto a replacement service provider, need to be thought through and included in the contract. From a purely logistical point of view there probably needs to be a transition period of some kind, analogous to that used at the start of the contract. Supplier contracts such as power, gas, chemicals, spares and so on will need to be novated, again either to the client or to a replacement utility service provider.

It is important to include clauses that oblige the service provider to co-operate during the transition back to providing the services in house, or to another out-sourcing contractor. Obligations such as the need to provide labour records. O&M records and manuals need to be included in the contract, although it should be recognised that for some of these, which may continue after the service provider has left site, it is reasonable to have to pay for.

PART IV

Energy Service Companies

INTRODUCTION

This chapter deals with the supply side of energy services and looks at the services that are currently on offer from providers, as well as potential future offerings. It also looks at the structure of the energy services industry in the UK.

The first section details the different types of services that are on offer. The second section deals with the types of service providers that are operating in the market. Finally, the chapter considers those companies that are linked to energy supply companies and looks at the motivations and drivers behind the growth in energy services being linked to energy supply.

All activities in the energy management process described in Chapter 4 could theoretically be outsourced or handled in house. The lack of resources, coupled with the pressures to focus on core business, mean that it may be attractive to outsource all of them to an external service provider. Most of the processes could competently be carried out by one or more service providers but very few, if any, of the current energy service companies (ESCOs) have the full range of capabilities to carry out all the energy management processes. There is overlap between potential service providers and different activities are outsourced to different service providers, therefore the management of interfaces is an important issue to consider. Any interface between the services of two service providers will be a potential cause of friction and argument.

Customers can and do also drive suppliers in certain directions. The energy outsourcing market is developing quickly and new services need to be developed to meet new challenges. As a customer, don't just accept what is on offer as the last resort. Go out and find an innovative service partner and develop new services that match your needs exactly. Innovative service providers need innovative customers.

CHAPTER 7

The Supply of Outsourced Energy Solutions

INTRODUCTION

This chapter looks at the energy services currently on offer in the UK. New service offerings are being developed by several providers and consequently this can only be seen as a snapshot of the UK market as of 2005–2006.

DEFINITIONS

Before looking at the offers that are available in the energy services outsourcing market it is necessary to be clear about definitions. Discussions about energy services and energy service companies (ESCOs) are often confused by lack of understanding of the basic definitions.

Bertoldi and Rezessi (2005) provide the following useful definitions:

> *An Energy Service Provider Company (ESPC) can supply energy efficiency equipment, heat, energy, Operations and Maintenance or Facilities Management but they do so for a fixed fee. They do not assume any on-going performance risks beyond that of normal warranties around product quality for example a 12 months defects period.*
>
> *An Energy Service Company (ESCO) can supply similar services to ESPCs but:*

- *they guarantee energy savings or the provision of services at lower costs*
- *their remuneration is directly tied to the level of savings/performance achieved*
- *ESCOs typically finance, or facilitate finance for projects they implement*
- *ESCOs retain an on-going operational role in verifying performance as well as often in O&M*
- *ESCOs are usually independent of any equipment vendor although some vendors offer ESCO like services*

The following sections review typical offerings from ESPCs and ESCOs.

ENERGY SERVICE PROVIDER COMPANY (ESPC) OFFERINGS

ENERGY PROCUREMENT CONSULTANTS

With the advent of the competitive market for power and gas that came about through the privatisation of the utilities in the 1980s, customers were for the first time able to pick and choose among energy suppliers. In each different jurisdiction the degree of competition and the level of demand at which competition applies varies, but there is an increasing liberalisation in nearly all markets around the world. With competition came the need for better procurement strategies on the part of customers and also came the rise of the procurement consultants. The services offered ranged from data collation through to advice on when to enter the market. The advent of the internet led to the rise of e-procurement and electronic auctions offering a full procurement and risk management service.

As the power and gas markets have evolved, the role of procurement consultants has changed. With the advent of flexible, transparent power products that allow better risk management, the requirements for the procurement consultants has changed to focus on risk management. As in other markets the diffusion of information through the internet has reduced the need for intermediaries.

TECHNICAL OR MANAGEMENT CONSULTING

Using consultants is the simplest form of energy services outsourcing. It is a way of accessing specific technical expertise that is not available within the organisation or obtaining an independent review of the situation. Consultants cover the following main areas:

- strategic reviews;
- designing new energy management programmes;
- energy surveys or audits;
- specifying and implementing monitoring and targeting systems (see below);
- implementation of energy management programmes;

- feasibility studies on specific opportunities;
- developing and implementing an energy awareness programme;
- advising on energy outsourcing.

As in all areas the expertise of consultants vary enormously and it is important to select an appropriately qualified and experienced consultant. In the UK the consultancy market has been affected by the intervention of the Carbon Trust, which supports energy surveys for a range of industrial and commercial organisations. Many of the initial surveys identified economically viable opportunities that were not followed up and having identified this problem; the Carbon Trust changed its programmes to focus more on detailed studies in specific technical areas aimed at promoting implementation of viable projects.

In recent years there has been an over-emphasis on specific technical support such as energy audits at the expense of strategic, overall programme design and management assignments. As well as the specific comments on the inadequacies of energy audits in Chapter 4, an audit or survey is only useful if it is within the context of a coherent, targeted programme. If it is not, it is highly unlikely to be useful and will end up as just another report with little or no follow-up.

Most consultancy is acquired on a fee basis, either on a daily rate or a fixed price basis, although some consultancies offer performance-based fee structures. Much of the success of engaging a consult comes from the quality of the brief. The brief should cover:

- why the project is being undertaken:
  - the background and reasons why the organisation has decided to undertake the project;
- what is required of the project:
  - aims and objectives of the project;
  - available budget;
- who is involved in the project:
  - internal resources;
  - other external resources;
  - supervision of project;

- when the project is to start and finish:
  - detailed work programme;
- where the project is to be completed:
  - scale and scope of project;
  - geographic location;
- how is the project to be carried out:
  - what the methodology is for collecting and analysing data;
  - the timeframe for its completion;
  - performance indicators to monitor and evaluate the project;
  - confidentiality of project and report;
  - managing the project and reporting requirements;
  - presentation of the final report (format, number of copies, need for a presentation);
  - ownership and copyright of final report and working papers.

For schemes in which the payment for consultancy work is linked to performance it is important to precisely specify the target outcome and the method of measurement.

MONITORING AND TARGETING SERVICES

As described in Chapter 4, monitoring and targeting (M&T) is a vital part of effective energy management. Without it you cannot measure progress against a target. Since the advent of M&T software packages in the early 1980s, in the UK most of the suppliers of software have also offered 'bureau services' in which they enter and process data and provide reports to the client, thus removing the need to employ additional staff. Bureau services usually include:

- Data entry. This can be by manual entry of invoice data (less common now), electronic data interchange (EDI), or direct collection of meter data, for example, in the UK half-hourly data
- Invoice validation and recovery. All invoices are checked and the bureau approves invoice payment. Given that historically invoicing errors can account for five per cent of the total energy bill, invoice validation can often pay for the entire M&T service.

- Finance reporting. Typical examples include actual costs versus budgets for different cost centres. In addition the bureau can also input cost data directly into accounts payable systems, thus avoiding the need for additional data entry.
- Billing of departments or sub-tenants. This is particularly important in manufacturing sites, multi-tenanted buildings or facilities like universities, which want to provide departmental cost centre information.
- Energy management. Energy performance reports comparing consumptions to budgets for the site as a whole or individual cost centres. For multi-premise organisations the bureau can also produce league tables and using exception reports focus attention on the areas where potential savings are highest.
- Energy procurement. The bureau service can feed historical consumption data directly into procurement process.
- Monitoring of water and effluent consumption and costs.

Most energy M&T bureaux can now provide reports over the internet via web browsers.

With the move towards flexible energy procurement and risk management, coupled with the increasing, and increasingly volatile, energy prices there is an increased need to improve the monitoring of energy information within the enterprise. An enterprise energy information system collects, analyses and distributes energy usage and cost information throughout the enterprise and can be used for energy supply procurement as well as energy demand purposes. As such it goes beyond M&T.

ENERGY SERVICE COMPANY (ESCO) OFFERINGS

The basic ESCO concept is shown in Figure 7.1. An ESCO develops and implements a project or projects that reduce the cost of energy and operations and maintenance (O&M) and takes a fee that is related to savings. The client achieves net savings after capital repayments from the start of the project.

DIFFERENT ESCO MODELS

Several different ESCO business models exist and this can cause confusion. Different countries have different dominant models and in different countries ESCO activities have different market penetrations in each sector.

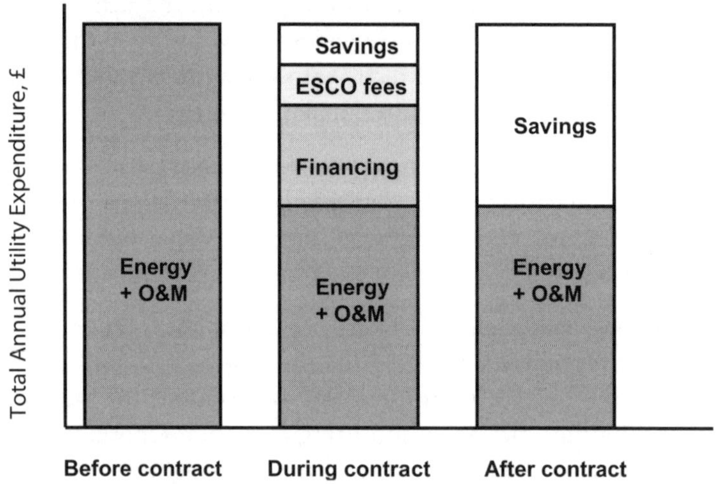

Figure 7.1 The basic ESCO concept

The main business models are:

- energy saving performance contracting (ESPC)
- build, own and operate (BOO)
- chauffage
- contract energy management (CEM)
- utility outsourcing.

ENERGY SAVING PERFORMANCE CONTRACTING (ESPC)

Energy saving performance contracting (ESPC) developed in the USA and is still the dominant model in that market. In the USA it is most often applied in commercial office buildings and public sector facilities. In the early days of ESPC, two models developed: shared savings and guaranteed savings. Two main variables determine the range of possibilities: the source of finance and the type of guarantee provided by the ESCO. The principal model in the USA (and elsewhere) has evolved to the 'guaranteed savings' model. In this model the ESCO guarantees the technical performance of the projects, that is, the level of energy savings that are to be achieved. The level of energy savings should be sufficient such that, at a defined energy price, the value of the savings will exceed the capital repayment costs for the project. Thus the client will be cash flow positive from the beginning of the project. In this arrangement the service provider is taking the technical performance risk and a financial institution financing the project is taking the credit risk.

Third-party financing evolved as an alternative to ESCO financing as with the latter the number of projects that the ESCO could finance would be limited by capital availability. Also, with ESCO financing, the ESCO has to take credit risk; a risk it is not necessarily skilled at, and its balance sheet would become overloaded with debt financing unless capital was coming from its own equity financing. See Figure 7.2 for an illustration of ESCO versus bank financing

The guaranteed savings, third-party finance model in which the guarantee is related to energy savings and a fixed or floor energy price is used, with finance being provided by a third-party financial institution, or sometimes the client themselves, came to be the dominant model.

Due to the international marketing of US-based ESCOs, aided by the USAID agency and multi lateral institutions such as the International Finance Corporation (IFC), this model of ESPC has been widely promoted throughout the world and examples can now be found in most developed, developing countries, central and eastern European (C&EE) and former Soviet Union (FSU) countries. However, it remains rare in the UK.

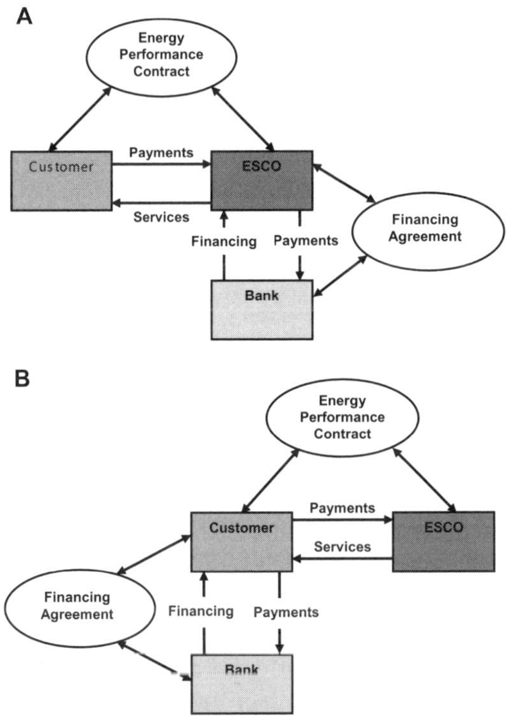

ESCO Borrowing

+ve	-ve
• Debt not on customers balance sheet	• ESCO borrowing limited
• Bank not involved in agreement between ESCO and customer	• Higher cost of capital
• Customer protected from financial performance of project	

Customer Borrowing

+ve	-ve
• Lower cost of capital – more money available for projects	• Debt on customer balance sheet
• ESCO balance sheet more like a service company than a bank	• Customer taking financial risk of project performance (mitigated by guarantee from ESCO)
• ESCO not taking credit risk of customer	
• Growth of ESCO not limited by capital availability	

Figure 7.2 ESCO versus bank financing (A) ESCO borrowing, (B) Bank financing

One variant of ESPC, which appears to be confined to the Canadian market, is First Out. In First Out the ESCO takes all the savings until the capital is repaid. This reduces the time that the ESCO is at risk but has the disadvantage from the client's view that there are no up-front savings. Savings do not come through until later in the project lifetime.

BUILD, OWN AND OPERATE (BOO)

In build, own and operate (BOO), the ESCO designs, builds and finances an energy asset for a specified contract term. The ESCO is bound by service level agreements (SLAs) and performance guarantees. For larger projects the ESCO can establish a special purpose vehicle (SPV), sometimes in partnership with the client, but for smaller projects this is not usually cost-effective. The BOO contract form is most often applied to are the larger energy conversion infrastructure assets such as boilers, energy centres, combined heat and power (CHP) systems, large refrigeration systems and possibly high-voltage and low-voltage (HV & LV) power distribution systems. It can be applied to larger compressed air installations as well as industrial gas installations. In new build situations it could also be applied to utility distribution systems as well as the main conversion equipment.

The revenue to the ESCO can either be as a straight fee or lease payment, or related to utility sales, for example, volume of steam taken by the client. In the latter case the ESCO will need to seek some form of 'take or pay' contract in order that it recovers the capital investment irrespective of the energy demands made by the client.

Figure 7.3 shows a typical BOO structure.

The major market for BOO in the United Kingdom (as well as other markets) in the 1990s was CHP. Much of the installed CHP capacity was constructed on a BOO basis, with many of the major energy suppliers establishing CHP subsidiaries to design, build, finance and operate these assets. Since 2001 the market for CHP has been stagnant despite the very real environmental benefits of CHP as the 'spark spread', the difference between gas and power prices, has been insufficient to justify the investment. Increased volatility in the gas market in late 2005 has added to the uncertainty for CHP constructors although there does in early 2006 there seems to be an upsurge in CHP feasibility studies, possibly driven by the carbon agenda rather than pure economics. Looking out beyond 2007 some forecasts show an improvement in the spark spread, which will help the economic case for new CHP plant.

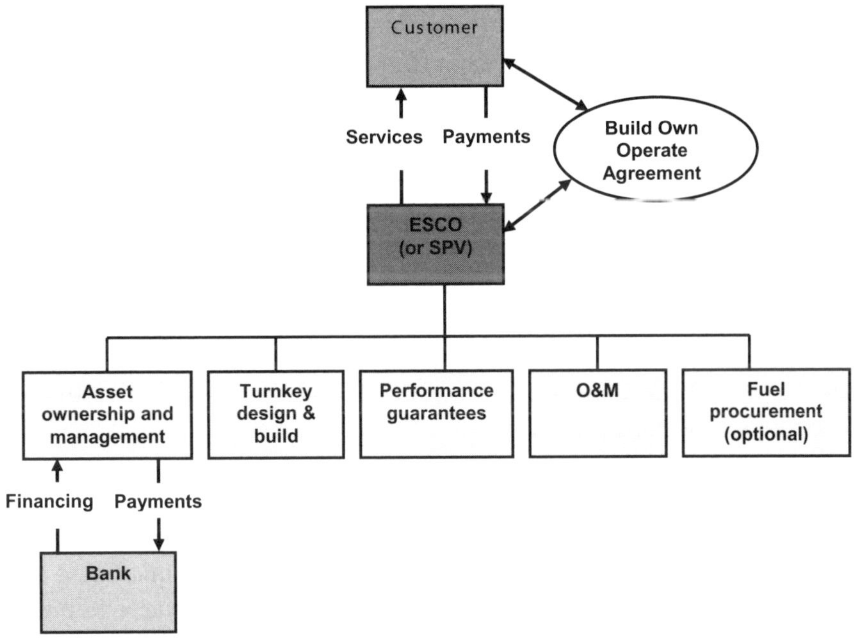

Figure 7.3 Typical build, own and operate structure

CHAUFFAGE

Chauffage is perhaps the oldest structure for energy services. In chauffage the ESCO takes responsibility for providing the utility service, usually heat (hence the term 'chauffage' in France, which has been adopted worldwide). Chauffage ESCOs either provide a guarantee of reduced costs (for example a five per cent reduction) or charge on a price per metre squared basis. Typical chauffage contracts are 20 to 30 years (with some on an even longer basis) and the ESCO usually takes on O&M. In fact it is usually in O&M that savings can be made. Typically the capital replacement risk still sits with the client, meaning if an asset such as a boiler fails completely its replacement is still the responsibility of the client. Chauffage dominates the market in France and French companies have exported the concept to other countries including the UK. The French Association of Energy Service Companies has developed a contract form that allows ESCOs to start out with simple chauffage and O&M, and move into capital replacement. A major problem with chauffage contracts is that they do not incentivise the ESCO to reduce energy demand. In fact, as they are selling heat, it is in the ESCO's interest to sell more heat and not less. A typical chauffage contract does not facilitate the ESCO in moving 'beyond the boiler house' to identify and implement energy demand reduction projects.

CONTRACT ENERGY MANAGEMENT (CEM)

Contract energy management (CEM) is the generic name applied to the energy services market in the UK. Much of the CEM activity is either BOO or chauffage with the emphasis on O&M. Most of the value created by CEM schemes has been from O&M and particularly by transferring O&M staff to the ESCO under Transfer of Undertakings (Protection of Employment) Regulations (TUPE). By transferring staff and de-manning boiler houses, with the required level of boiler automation, and putting 'men in vans' looking after several boiler houses in a local area, ESCOs could provide the same level of service at lower costs. There were some gains in energy efficiency of conversion equipment when CEM/BOO was used to replace boiler plant but very little investment in projects to reduce energy demand.

As with chauffage the ESCO does not typically assume any capital replacement risk or take any interest in reducing energy demand; in fact, in many of the early CEM deals the ESCO did not even assume O&M risks. Clients often bear the risk of repairs and replacements and have to pay the costs of these as they are needed. These 'first generation' CEM deals are still common but are increasingly coming under question as pressures to reduce energy costs increase. In many cases further de-manning exercises, or savings from reducing O&M costs, are not possible and hence attention has to switch towards energy saving projects to create extra value. It will be interesting to see if the dominant CEM companies in the UK can make this transition, which requires different skill sets and a different approach.

MULTI-UTILITY OUTSOURCING

Multi-utility outsourcing appeared as a new form in 2001 with the innovative contract between Enron Energy Services and Diageo for the Guinness Park Royal Brewery in London. It covered all utilities used on the client's site including water, effluent and industrial gases. As with chauffage, in this contract form the ESCO sells the client utilities (typically steam, refrigeration, compressed air, treated water, HV power, industrial gases) by volume (pounds sterling per tonne and so on) according to SLAs covering factors such as quality (temperatures, pressures and so on) and availability. The contracts involved transferring O&M staff to the ESCO. The deals also included the facility for shared savings projects in which both the client and the ESCO could benefit. The first important innovation here was that these projects could be identified by either party and could be 'either side of the fence', that is in energy supply or energy demand. This was an attempt to align the objectives of the two sides, which in traditional CEM were not, and still often are not, aligned. The second innovation was that the contract required the ESCO to purchase all utility assets from the main incoming meters to the 'last valve or switch' before the process. This was novel in that ESCOs had never monetised existing utility assets before. It gave the client an immediate injection of cash and reduced the value of assets employed, important if management performance is measured in terms of return on capital employed (ROCE). The third innovation was that the ESCO identified major projects to reduce costs up-front and undertook to implement them within a certain period and furthermore, they were incentivised to do so because the prices charged for each utility took into account the savings resulting from the projects. This arrangement transferred project performance risk and project completion risk firmly to the ESCO.

After the collapse of Enron in Europe, Diageo tendered the project and RWE Solutions UK Limited (now part of RWE npower) won the tender and both parties signed a deal similar to the original in 2002. The concept was later extended to two other Diageo breweries, Dundalk and St. James Gate in Dublin. This form of multi-utility outsourcing represents one end of a spectrum of complexity and risk management for the ESCO. Such deals have to have a long term (typically 15 years), have high transaction costs, and require a high degree of intellectual capacity to develop and manage. They are strategic partnerships that resemble structured finance deals in the banking world and like them have a high transaction cost, particularly in legal fees.

Figure 7.4 shows the structure of a typical multi-utility outsourcing project.

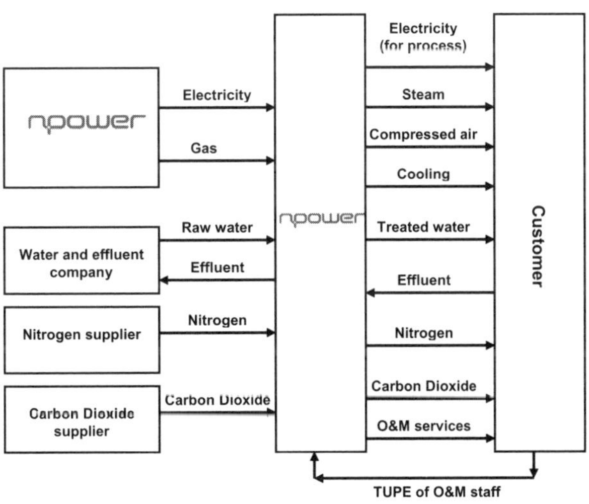

Source: *RWE npower*

Figure 7.4 Structure of multi-utility outsourcing projects

SOURCES OF VALUE

Within the energy operations of any company there are four main value drivers: energy prices, energy demand, O&M costs and capital. Different types of ESCO deal focus on different value drivers.

Figure 7.5 illustrates the sources of value and maps where typical deal structures fall in terms of the two axes – supply side to demand side, and O&M to energy. In this context supply side means the conversion equipment such as boilers and compressors. As can be seen, the different deal structures overlap. To date the ESCO market in the UK has mostly addressed the supply side/O&M quadrant of the equation and not the demand side/energy quadrant. As energy prices have increased, the potential market in this quadrant has grown but there has been little or no supply offers into this quadrant as most UK ESCOs have different skill sets (focused on O&M) and have considerable investment in their existing deals and structures.

PARTIAL VERSUS TOTAL OUTSOURCING

As already discussed, energy management outsourcing has to date been partial, with few if any examples of total energy management outsourcing having occurred. RWE npower has gone further than any other supplier to provide a total outsourced solution and continues to innovate with major customers in the areas of risk and energy management. Considerable value remains to be created by fully developing new total energy outsourcing models.

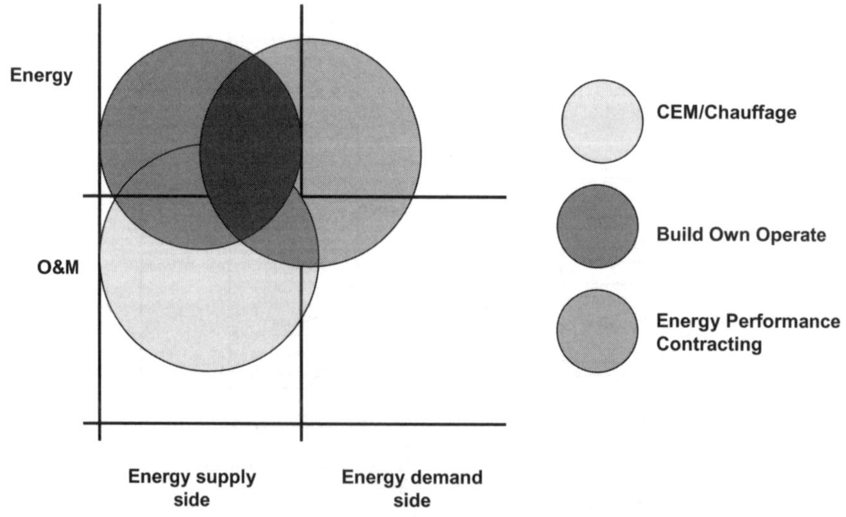

Figure 7.5 Sources of value from different deal structures

HISTORY OF THE UK ESCO MARKET

Box 7.1 shows the major events in the development of the UK ESCO market.

Box 7.1 Major events in the history of the UK ESCO market

Year	Event
1966	Associated Heat Services (AHS) formed by National Coal Board to provide outsourced management of boiler houses
1980	First ESCO based on energy performance contracting model (Utility Management Company)
1980–1982	Rapid growth in coal fired leased, packaged boiler houses assisted by coal conversion grant (AHS install 36 in one year)
1982	Emstar formed by Shell
1983	BP Energy formed
1986	Local authority financing rules changed to allow local authorities to use external financing
1989	AHS and Emstar merge
1990	Dalkia takes over AHS Emstar
1990s	Rapid growth in packaged gas turbine CHP market
1999-2001	Enron Energy Services active in UK
2000	Cofathec takeover of Heatsave
2002	RWE Solutions enters UK market
2002	Elyo purchases BP Energy

The market began in 1966 with the formation of Associated Heat Services (AHS) by the National Coal Board (NCB), the nationalised integrated coal company. In 1980 a small start-up company, Utility Management Company, became the first UK company to undertake energy saving performance contracts, working mainly in the public sector. In the early 1980s the government was promoting the use of coal for industrial boiler houses and there was a rapid growth in the BOO of packaged coal boiler houses in 'energy pods'. In 1982 Shell decided to enter the market and set up Emstar, the first large-scale ESCO. Emstar entered the industrial market and tried to work in the public sector market, something that was inhibited by Treasury guidelines on local authority financing that were not changed until 1986. In the mean time BP had established BP Energy as a competitor to Emstar. At the end of the 1980s AHS and Emstar merged and shortly after were taken over by Dalkia, a subsidiary of two French Companies, Electricité de France (EDF) and Veolia Environnement. As already noted above, in the 1990s the growth market was CHP. In 1999 Enron Energy Services (EES) entered the UK market and prior to the collapse of Enron in 2001 was making waves in the market through its innovative contract with Diageo and another contract on a different basis with the retailer Sainsbury's. After the collapse of Enron a small team from EES (including the author) joined RWE Solutions and built upon the experience gained by winning tenders to replace the Diageo and Sainsbury's projects and used this base to establish a new force in the ESCO market. In 2002 BP Energy exited the market by selling its CHP portfolio to Elyo, a subsidiary of the French utility Suez Lyonnaise des Eaux.

UK MARKET SIZE ESTIMATES

It is difficult to measure the ESCO market size precisely due to the variety of activities being undertaken but most estimates put it at between £500 and £750 million per annum. As noted above, most of this activity is CEM/BOO/chauffage, which has only had a marginal effect on energy efficiency. The greatest impact on energy efficiency would have been due to the CHP installations. The market focussed on CEM/BOO/chauffage is mature and competitive although there is growth in the public sector. Clients in the industrial sector are finding it harder to extract meaningful value from these kinds of contracts and looking for new forms of outsourcing which focus on reduction of total energy costs. An estimate of the potential market for energy efficiency and carbon management focused services is £800 million per annum.

COMPANY PROFILES

Profiles of the major companies operating in the UK market are shown in Boxes 7.2 to 7.7.

Box 7.7 gives information on other energy services companies in the UK market. In addition, a number of other market participants exist. Examples include Chalmor, a manufacturer of heating and lighting controls, who have pioneered a 'pay as you save' scheme in which they will provide

Box 7.2 RWE npower

Ownership: RWE AG (Germany)

Main models: BOO, utility outsourcing and performance consulting, total energy outsourcing

npower Energy Solutions: Provides a range of energy services across industrial, commercial and public sectors. Services include:

- supply of electricity and gas on fixed and flexible contracts;
- risk management
- identification, development, financing and O&M of new embedded utility infrastructure;
- identification, development and financing of energy efficiency and carbon management projects including CHP and renewable energy systems.

Box 7.3 Dalkia

Ownership: 66 per cent Veolia Environnement (France)/34 per cent EDF (France)

Main models: CEM/O&M

Dalkia Utility Services: Provides energy, utilities and technical facilities management to industrial, hospital, community and public sector clients.

Dalkia Energy and Technical Services: Provides an integrated service management approach combining energy management, M&E services and a range of related services including fire protection, environment management and energy bureau.

Box 7.4 Elyo

Ultimate parent: Suez (France)

Main model: CEM/O&M

Elyo Industrial: Integrated energy and utility management services, in addition to technical facilities services, for the industrial sector.

Elyo Services: Buildings and operations management. Owns Axima Building Services (design, supply, installation and commissioning of building services) and IGS Services Ltd.

finance to install their controls for the schools market. ABB, the international manufacturer of electric motors and variable speed drives (VSDs) also provide various financing options in which savings exceed the capital repayments as do Johnson Controls in the public sector market. TM Resource is a long established consultancy activity in the industrial market that charges fees based on the savings achieved.

Box 7.5 Cofathec

Ultimate parent: Gaz de France (France)

Main models: CEM/O&M

Provides energy management and building services. Includes, for example, PFI contracts (i) for the provision of full energy services to nine prisons and (ii) to design, build and operate a civic centre for Trafford Metropolitan Borough Council (including the provision of full FM and energy management for 25 years).

Box 7.6 E.On

Ultimate parent: E.On AG (Germany)

Main models: BOO and CHP

E.On UK CHP Ltd (formerly Powergen CHP Ltd): Sells energy and related services including the construction and operation of CHP. The company has 'no current plans to initiate any new CHP development projects' (taken from *Report of the Directors for the year ended 31 December 2004*).

Box 7.7 Other UK ESCOs

Company	Ultimate parent	Notes
ENER-G	ENER-G plc/Ursa Major Management Ltd	Mainly small-scale CHP, moving into controls and ESCO style projects in public sector
Utilicom	IH International SA (Luxembourg)	District heating – subsidiaries include Southampton Geothermal Heating Co., Bloomsbury Heat & Power and Gower Street Heat & Power
United Utilities Energy Services	United Utilities plc	Principle activity is to develop 'shared savings' projects with customers benefiting from energy efficiency measures without using their own capital
SSE Energy Services	Scottish and Southern Energy plc	Established 2004 to provide 'beyond the meter' services. Initial focus on public sector organisations
MCL Energy	[none]	CEM and O&M. Plants served provide steam, hot water and CHP

BARRIERS TO MARKET GROWTH

The major barriers to the growth of the ESCO market are:

- transaction costs;
- negative perceptions of outsourcing;
- perceived risks;
- M&V of savings;
- difficulties of financing energy efficiency assets;
- lack of supply options (with most supply options emphasising O&M).

In energy intensive process industries such as chemicals, cement and glass, the opportunities for ESCOs to be involved in demand-side projects are limited because energy is so intimately linked to the process that the client will often have greater expertise than ESCOs and only major process innovation will have a significant impact on energy use. In these sectors, stand-alone supply-side offerings such as boilers or CHP are more likely to be acceptable. Demand-side ESCO offerings are more likely to be appropriate in less energy intensive sectors such as food and drink and pharmaceuticals.

A major barrier in the commercial building sector is the tenant-landlord problem in which the landlord only has responsibility for energy use in the common areas such as corridors and lobbies whilst tenants have the largest effect on energy use in their areas through selection and use of heating, ventilating and air conditioning (HVAC) equipment but pay energy bills indirectly through service charges. The wider use of sub-metering and billing will help to reduce this problem.

In the public sector there are sometimes still negative perceptions about outsourcing based around ideas such as 'the contractor is taking all the savings' or 'how can we measure whether the client is performing?' Public private partnerships (PPPs) have not addressed the energy efficiency issue until recently and many opportunities to build low energy buildings in this sector have been missed. To counter the measurement issue it is essential to build in M&V and to use the International Protocol on Measurement and Verification (IPMV) as a standard.

In the small and medium enterprise (SME) sector the barriers are high credit risks, high transaction costs relative to the size of the deal, and a lack of suitable 'packaged' products built around standard technologies. The bespoke approach that is taken in larger projects cannot be applied in the SME market. What is needed in this sector is standardised packages incorporating specific technologies such as lighting controls, heating controls, voltage regulation or variable speed drives (VSDs).

MARKET TRENDS

The biggest trend that is changing the market in the UK and other countries is the involvement of utilities in the energy services market. Most energy service companies (ESCOs) were formed from either engineering consultancy companies or companies with energy efficiency products to sell. They had little to do with energy suppliers or utilities unless their scope included operating M&T or procurement-related services and then they were mainly dealing with utilities' sales teams or their billing and accounts sections, most often resolving billing errors. In today's outsourcing markets, however, utilities are directly or indirectly, major players. In the USA the number of utility-owned ESCOs has grown dramatically over recent years. In the UK three of the largest European utilities, RWE, EDF (Electricité de France) and GDF (Gaz de France), all own energy service companies. The degree of integration between the energy supply arms and the energy service companies varies greatly, with RWE npower having the most integrated energy services offerings. These utilities also own energy services companies in other markets including in some cases, markets where they are unable to supply energy at the moment, such as countries in central and eastern Europe.

There are several factors behind this trend, including:

- EU regulations are promoting the shift from energy companies to energy service companies;
- an energy service company can add value for energy suppliers;
- an energy service company can differentiate an energy supplier in a competitive energy market;
- owning an ESCO can contribute to improving the environmental and CSR reputation of a utility;
- managing customer demand by investing in demand-side projects and energy management can reduce the need for an energy company to invest in supply infrastructure such as cables and transformers;

- investing in managing demand can open up new profitable opportunities such as load management, obtaining renewable obligation certificates;
- the fact that ultimately energy buyers do not want to buy energy: they want to buy the services that energy can produce such as light, heat and refrigeration;
- customer demand to reduce or at least stabilise energy bills at a time when energy prices have risen sharply;
- in the UK the EU ETS has been configured by the government such that most of the pain of reducing emissions falls on the generators of power; one option they may have available to them to assist in reducing overall emissions from their portfolio is to invest in energy efficiency and load management or to buy carbon credits;
- energy services are seen as a relationship platform for selling energy.

The ownership of ESCOs by utilities has a number of advantages, which can be divided into advantages for the utility, the ESCO and the customer. These include:

Access to greater financial resources. Utilities tend to be large enterprises in all markets and the investments in energy service deals can be relatively small compared to the investments in generating assets. (Conversely this may also represent a barrier as (a) the investment opportunities may be too small and (b) people used to managing and investing in generation assets may not understand energy efficiency and end-use investments.)

Providing a 'one-stop shop' for client. This is viewed in different ways but generally there is a move within the procurement field to reduce the number of suppliers and move towards fewer strategic partners.

The ability to combine the technical world with the trading world. The best example of the ESCO acting as a bridge between the physical world and the virtual world of trading energy is load management. Loads to be managed have to be 'despatchable' and quite large (above 5 MW) to have value. An ESCO, working for example for a large retailer, could load manage HVAC fans or chillers through a building management system (BMS), which could be controlled from the traders screen at times when the value of load management is high. Each HVAC fan in itself may be small, 40 kW, for but when they are multiplied up by 1000 (the number installed in some 500 stores for example), the potential load shedding is 40 MW, a size that has value to the grid. This kind of project can create significant value through being able to sell the reduced load to the grid or the power saved to other customers prepared to pay a premium at that point in time. A similar example would be bringing on multiple standby generators to produce power during a spike in prices.

Moving into the clients' sites and controlling, where possible, or at least influencing, energy demand gives the utility a greater understanding and knowledge of what determines actual loads. The ability to accurately predict loads is essential to a modern utility as there are penalties for getting the predictions wrong. The benefits are in being able to price power (and gas) more reliably, which should give a marketing advantage, and reducing risks for the utility.

A utility-owned ESCO will be able to access other resources within the utility, for example, the ability to trade carbon through the utility's trading floor. Integrating services and the sale of carbon saved in customers' sites can generate an additional income stream that can benefit both the client and the ESCO.

COMBINING ENERGY SUPPLY AND ENERGY DEMAND SERVICES

There is a trend to incorporate energy supply and energy demand services into one integrated package. In the US companies such as Cinergy (now part of Duke Energy) offer a 'one-stop shop' for both energy supply and energy demand while in the UK RWE npower is the leader in this area.

For the supplier, the benefits of linking energy supply with energy services can be split into five categories: improved profitability, differentiation, compliance, improved client relationship and reduced risk.

Within 'improved profitability' the following benefits can be realised:

- Energy services typically have higher profitability than commodity supply. In the UK commodity supply margins are typically one to two per cent, whereas energy services contracts typically have much higher margins.
- Offering energy services can bring more clients to the energy supplier.
- Energy services take the energy supplier further into the value chain.

- The energy services market is growing.
- By combining energy supply and energy services deals energy suppliers can obtain far longer energy supply deals, in some cases five, ten or even fifteen years.

Within 'differentiation' the following benefits can be realised:

- Offering integrated energy services is a differentiator.
- Enhanced brand image by increasing visibility of environmentally sound practices and investments.

Within 'compliance' the following benefits can be realised:

- All utilities are facing tougher environmental emissions targets and energy services can help to offset carbon emissions from generation.
- In the UK all generators have to produce a certain percentage of power form renewable energy and can purchase or sell ROCs, which have a value in the market. There is also currently a shortage of ROCs as renewable energy developers struggle to bring projects on-stream in sufficient quantity and in sufficient time to meet demands from generators and customers. Investing in energy services, which can include on-site renewable generation, would help to increase the availability of ROCs.

Within 'improved client relationship' the following benefits can be achieved:

- Billing issues can be resolved through energy management techniques such as automatic M&T (aM&T).
- Moving up the value chain and decision chain within organisations leads to a higher profile.
- It moves the supplier away from being a simple commodity supplier and towards becoming a strategic, long-term partner.

Within 'reduced risk' the following benefits can be achieved:

- By understanding customer demand better, and in some cases being able to control customer demand through load management strategies (pre-agreed with clients) such as switching off HVAC fans for one hour at periods of high demand, utilities will be able to reduce their exposure to market risks inherent in the electricity market.

The view from the customer

The potential benefits to the customer of 'bundling' supply and demand services fall into four areas: improving profitability, reducing risk, accessing resources and image.

Within 'improving profitability' the following benefits could be achieved:

- reduced total energy costs;
- client is allowed to focus on core business;
- reduces investment needs;
- sale of assets can bring in funds;
- successful energy management can produce an additional revenue streams through tradable outputs such as ROCs or emissions trading under schemes such as the EU ETS or JI CDM;
- combining energy supply and energy demand into one procurement activity can lead to administrative and procurement savings;
- billing issues can be resolved by 'inserting' an energy service provider between the commodity supplier and the customer, or by application of 'smart metering' and monitoring technologies such as a M&T.

Within 'reducing risk' the following benefits could be achieved:

- reduced exposure to cost variations;
- passing energy conversion risk onto a service provider;
- taking advantage of any guarantees on savings or performance;
- operational risks can be passed onto a service provider;
- higher reliability bought about by improved O&M and/or increased capital investment in new utility assets.

Within 'accessing resources' the following benefits can be achieved:

- added resources to focus on demand management;

- access to specialist technical resources independent of equipment vendors;
- access to market trading expertise for energy commodities and emissions;
- access to greater investment capital;
- access to new ideas, techniques and technologies;
- access to influence wielded by large utility groups on industry-government working groups and so on.

Within 'image' the following benefits are possible:

- being seen to follow a sustainable energy policy – 'putting words into actions';
- meeting government pressures and targets on emissions;
- image enhancement from being seen to reduce environmental impacts.

Given the advantages to both parties, as well as the wider benefits to society bought about by introducing true energy services, the combination of energy supply and demand services will continue to grow.

CHAPTER 8

Future Trends for Energy Service Companies

I believe we can accomplish great and profitable things within a new conceptual framework – one that values our legacy, honors diversity, and feeds ecosystems and societies... It is time for designs that are creative, abundant, prosperous, and intelligent from the start.

William McDonough

One of the greatest pains to human nature is the pain of a new idea. It makes you think that after all, your favourite notions may be wrong, your firmest beliefs ill-founded... Naturally, therefore, common men hate a new idea, and are disposed more or less to ill-treat the original man who brings it.

Walter Bagehot, *Physics and Politics*

We shape our buildings and afterwards our buildings shape our lives.

Winston Spencer Churchill

INTRODUCTION

The market for outsourced energy services is evolving quickly in many markets. In the USA it was estimated in 2000 that energy service companies (ESCOs) have a revenue of $2 billion per annum and that the industry was growing at 24 per cent per annum (Goldman et al., 2002) while across Europe the market is estimated at €5 billion a year (Bertoldi and Rezessy, 2005).

This chapter looks at the directions and some specific areas in which energy services companies working with customers may develop new services.

INFORMATION SYSTEMS

As we have seen, monitoring and targeting (M&T) is a powerful management technique that is central to successful energy management. In the UK and elsewhere there have been recent advances in automatic M&T (usually known as aM&T) that offer the potential to make large savings. Automatic M&T involves the automatic collection of data from utility meters feeding directly into M&T software to produce the required reports. There is a good business case for applying aM&T to all organisations and the reducing price of data collection and transmission make it viable even for smaller organisations.

At the same time as aM&T is evolving, the energy supply industry is moving towards automatic meter reading (AMR or so-called 'smart metering'). Smart metering can bring about many benefits to utility companies including:

- reduced administration costs when customers move out of and in to new premises;
- reduced administration costs from eliminating manual meter reading and reducing billing errors;
- the potential to use two-way communication to exercise load management, that is control plant and equipment at the customer's facility;
- a quicker response when power fails.

Smart metering also offers the prospect of the customer gaining by being able to respond to peaks in demand and differential tariffs. Smart meters can be tied to building management systems and other control systems to automatically reduce demand at times of peak costs. Examples of this may include switching off or reducing the flow of ventilation fans in a mechanical ventilation system for 30 minutes, something that is unlikely to be noticed by most building occupants. The use of aM&T for self-billing would give an energy supplier more accurate information and potentially, better cash flow, as well as reduced administrative costs in resolving billing errors. Billing errors are a major source of customer dissatisfaction with utilities and, according to one survey, energy managers spend 70 per cent of their time resolving billing errors (Martin, 2005).

There is clearly convergence between aM&T and AMR. The customer wants useful information at the least cost. Combining the two techniques as part of an energy supply deal offers scope for cost reduction for both energy customer and energy supplier.

The changes in energy procurement, with more customers switching to flexible purchasing, have bought new demands for energy information for risk management purposes. Very often information for energy supply and energy demand sits in different systems, as the two functions are separate. There is a need for integrated enterprise energy information systems that pull together all the information required to manage energy supply and energy demand issues.

INTELLIGENT AGENTS

The electricity market is moving towards dynamic pricing in which prices change according to demand levels and real-time control of load would be used to minimise expenditure. Intelligent software agents have been developed that can control the operation of distributed generators and electrical grids (Gibson, Pratt and Patterson, 2001). At the consumer level intelligent buildings, industrial facilities with distributed energy management systems (SCADA), and even homes could use intelligent software agents (multiagent systems or 'Homebots') to optimise energy consumption. Some modelling shows that energy savings of 40 per cent could be achieved compared to conventional ways of operating a building (Davidsson and Boman, 2000).

TECHNOLOGIES

EMBEDDED OR DISTRIBUTED GENERATION

For some time environmentalists and others have promoted the idea of supplying energy needs from small-scale, embedded generating plants, using technologies such as combined heat and power (CHP or cogeneration) and trigeneration (simultaneous generation of power, heat and cooling) and renewables. This approach offers the potential to save energy through using CHP/trigeneration and reducing transmission and distribution losses in the power grid. The emergence of new technologies such as small-scale gas turbines with a capacity in the hundreds of kilowatts (rather than the megawatt-scale machines), small-scale embedded renewable energy systems and the emergence of fuel cells, plus the growing carbon agenda has bought the idea of embedded generation firmly back into focus.

The deployment of embedded generation technologies fits well with the ESCO model. A number of projects involving public-private joint ventures and the use of cogeneration/trigeneration with distribution of power over private wire networks have emerged in recent years, most notably in Woking, England. The sale of power to residential and commercial customers over private wire networks can boost the return on these projects compared to selling power into the grid. This is an area that offers potential growth for suitably equipped

ESCOs and could be a disruptive technology in the sense that it could start the break-up of large scale grid based systems if proven to be profitable.

As the energy market moves towards smaller-scale, distributed energy sources, and the building market moves towards more sustainable, energy efficient design, new opportunities will open up for ESCOs, particularly those linked to energy supply companies. There is now the prospect that many new buildings, even domestic properties, could now become net energy producers using emerging technologies such as building integrated photovoltaics (BIPV), on-site wind and micro-generation. Energy supplier owned ESCOs assisting in the design, specification, funding, construction and O&M of such facilities would be buying access to green power which is in increasing demand and long-term business relationships. Many of these technologies such as BIPV have marginal capital costs compared to using conventional roofing or building facade glazing. An ESCO could provide the marginal capital cost of adding renewable energy systems to new buildings. This approach is likely to be increasingly attractive to all parties as local authorities seek to impose sustainability targets for example the ten percent of energy to come from renewables required by the new building regulations, on new-build developments.

The growth of demand for green electricity and the value of renewable obligation certificates has created a demand for on-site, embedded renewable energy systems, particularly wind power. A number of high-profile industrial sites including Ford at Dagenham and Nissan in Sunderland have installed multi-megawatt wind turbines. With the increase in power prices in 2005–2006 the payback period on these installations can be as low as three years. The major issues are siting and gaining planning permission, but a number of service providers will now undertake the evaluation and development of these kinds of projects at little or no risk to the customer. As well as wind power, opportunities exist for hydro-power (often refurbishing or replacing old disused equipment on mill races) and solar power (BIPVs). The economics of solar power remain questionable unless they are incorporated as marginal cost items into building refurbishment or construction projects as cladding or facades. Several approaches currently being developed by start-up companies backed by venture capitalists, including the use of biological type systems and nano-technology, offer the prospect that there will be a disruptive technology that makes solar power extremely attractive. Thermal solar energy for pre-heating hot water for use in central heating or domestic hot water systems is a proven technology with reasonable returns at current energy prices that is largely ignored by government policy and commentators. In terms of payback and tonnes of carbon saved per pound of capital expenditure, solar thermal remains one of the most attractive renewable energy options, even in the UK.

Walt Patterson (1998) has pointed out that commercial and domestic consumers require increasing amounts of low-voltage direct current (DC) power to run devices such as computers, telephones and so on. Low-voltage DC power from batteries is expensive and wiring a building with low-voltage DC power, as well as the normal higher-voltage alternating current (AC), may make solar PV technology more attractive economically as it removes the need for conversion from DC to AC and transforming the voltage to 220 volts. In the same way that companies are prepared to pay more for high-reliability power because the cost of power failure to a data centre can be huge, they may pay more for low-voltage DC power. Provision of low-voltage DC power generated by PV could be a future opportunity for ESCOs.

EMERGING TECHNOLOGIES

A number of other emerging technologies that can reduce energy use and O&M costs are worth noting. These include:

- advanced radiant heating
- orifice steam traps
- ground source heat pumps (GSHP)
- BIPVs
- building integrated wind power
- light emitting diode (LED) lighting.

LED lighting offers the prospect of very significant savings in lighting as well as longer lamp life. Currently LED lighting is viable for specialised colour applications but not for general white lighting.

INFORMATION-BASED ENGINEERING DESIGN

The design and specification of energy systems is based upon design codes which are (a) inherently conservative and (b) based on the optimisation of units within a system rather than the entire system. The conservatism of design methods is understandable at one level as consulting engineers have to be conservative in safety and performance, however it does lead engineers to systemically oversize capital investment in facilities. This makes sense because they can be punished for putting in insufficient capacity; no one knows if they put in too much and the fee structures for consulting engineers are traditionally based on a percentage of the capital cost.

Normal mechanical systems are usually sized as follows:

- An engineer estimates the peak running load using a rule of thumb, what worked on the last job or an established code of practice.
- These will give a range of estimates, and to avoid blame in the event that something goes wrong, the largest value will be picked.
- A safety factor is added to the estimate (probably by the engineer and their boss).
- A level of redundancy is selected, for example n + 1, n + 2 and so on.

In an information-based design process, empirical data is collected in order to understand the load that is being sized for. For example, on a cooling system measurements can be taken on chilled water supply and return temperatures, flow rates and outside air enthalpy to give tonnes of cooling and a correlation with the major factor driving cooling variance. The sizing process would then be as follows:

- The peak load is derived from the empirical measurements. As there is engineering uncertainty in the measurements an appropriate upper limit is selected.
- The same safety factor is added to the peak load although this is a smaller number because the initial data collected has given the engineer a better understanding of the design parameters.
- The same level of redundancy is added.

The net result is a system sized some 35 per cent smaller than via the normal design process, thus directly reducing capital cost. Operating costs are lowered through:

- the plant running closer to its design loads and within its operating envelope for a larger fraction of the time (as opposed to severe part load) with improved plant efficiency;
- reducing ancillary loads;
- reducing maintenance due to selection of smaller plant and plant operating consistently within design loads.

Figure 8.1 compares normal equipment sizing with information-based engineering sizing.

A related technique is holistic engineering, which aims to optimise the performance of the entire system rather than individual components. Traditional services and process engineering seeks to optimise individual components such as pumps or boilers rather than the overall system (see Box 4.10).

The use of data-driven engineering and holistic optimisation techniques offers the potential to significantly reduce energy consumption, as well as capital cost, from the design and installation of new equipment and buildings. The 2006 building regulations will help drive the use of these techniques, but in order to force consultants to design systems in this way, educated clients have to push for radical change and something that is not 'more of the same'. This includes writing tough performance-based specifications to challenge creative engineers and reversing the normal incentives. Another factor that helps to perpetuate the existing engineering design approaches and leads to over specification is the system of paying consulting engineers a fee based on a percentage of capital expenditure. Creatively challenging this system, for example, rewarding the design engineers according to how efficient the final installation is in practice could lead to significant benefits. Any customers, or indeed ESCOs, using design engineers for new systems would be advised to consider this approach in preference to the standard fee agreements based on project capital cost. Often in business pressures to keep up production, or get a new production line in place, mitigate against this as it is easier to go out and

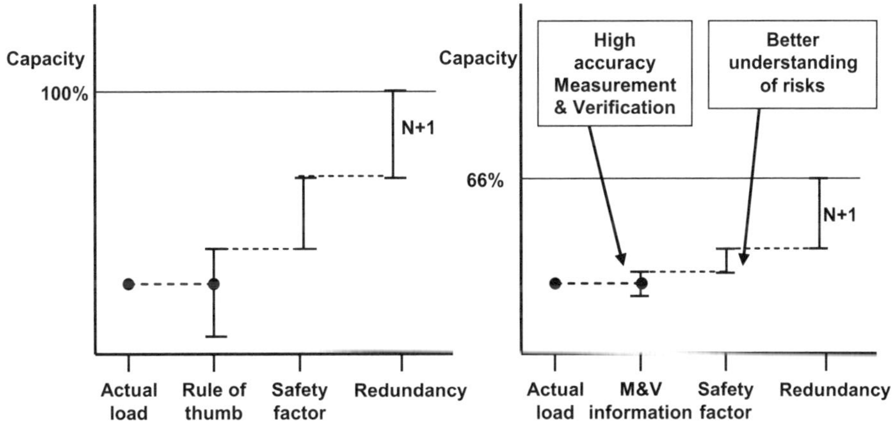

Figure 8.1 Normal equipment sizing versus information-based engineering sizing

buy the same kind of product in the same way. Clients must build in energy objectives at the beginning of any project that involves new HVAC or process utility plant (whether new build or retro-fit).

If these design techniques are applied energy savings of 30 to 50 per cent are very possible, against the norms of 5 to 15 per cent (RMI, 2006). The prospect of achieving this level of savings from projects would greatly increase the attractiveness of outsourced energy solutions and increase margins for ESCOs.

ENVIRONMENTAL SERVICES

With the increasing importance of the carbon agenda, there is a need for companies to access carbon trading services. The value of energy saving projects can be enhanced by the value of carbon credits, which can be sold on the EU ETS market. A holistic view of the entire value chain including carbon credits will add value that can be shared between customers and ESCOs.

As well as providing energy efficiency services there is potential for ESCOs to move into environmental projects as well as energy projects. There are now many examples which demonstrate that environmental projects can actually produce a positive economic benefit whereas previously they were seen as a cost burden. Companies such as 3M with its '3P Program', Dow and Chevron have consistently demonstrated profitable economic returns over many years from environmental projects (Hart, 2005). Implementing these kinds of projects in a contract structure similar to that used for energy service projects could be a market opportunity.

The cutting edge of sustainable development is for industrial facilities and buildings to be redeveloped in an integrated, sustainable way. One of the best examples of this is the Ford Red River plant in Detroit, an icon of the first industrial revolution, which underwent a $2 billion renovation to turn it into a cutting-edge example of sustainable development. Redevelopment of the site included a green, living roof, porous parking lots to aid water run-off, a biological water purification plant and daylighting (McDonough and Braungart, 2002). Benefits included reduced capital cost compared to conventional systems, reduced operations cost, greatly reduced environmental impact, increased bio-diversity on the site and improved worker satisfaction, as well as less tangible public relations and corporate social responsibility benefits. Using this kind of approach there may be opportunities for 'green utility service companies' to use a combination of sustainable architecture and engineering to develop and fund new or refurbished industrial facilities that achieve these kinds of benefits.

ALTERNATIVE FINANCING MECHANISMS AND STRUCTURES

An area that has not been explored much to date is alternative financing mechanisms and structures. One idea is to repay capital used for energy saving projects, through an increase in the power price and specifically an increase in distribution use of system (DUOS) and transmission use of system (TUOS) charges. This links payment to specific meter points and in the case of the domestic market assures payment over a long period when households may move before the capital is repaid. Similar ideas may be attractive in the industrial and commercial markets, particularly in the small and medium enterprise (SME) market. Paying a higher unit price is attractive if consumption is reduced enough to bring about a lower total cost.

The transaction costs of establishing an ESCO type project are high. One approach to reducing transaction costs is to put a portfolio of facilities into a framework agreement with an ESCO that allows the ESCO to implement projects across the portfolio. This approach has been used widely in US government federal facilities (through the so-called 'super ESCO' idea in which framework contracts are let that cover groups of federal facilities), and in other markets, notably Berlin. To date this approach has not been used in the UK. Large, multi-site energy users, particularly those in the public sector, should consider how similar arrangements could be put in place (while keeping in line with government and EU procurement rules). Obvious possibilities for this approach include the National Health Service, the Ministry of Defence, and local authorities, either individually or in consortia. A variation on this would be to assemble consortia of customers within a geographical area or a sector for example hospitality to allow them to access the resources of an ESCO.

PUBLIC PRIVATE PARTNERSHIPS (PPP)

In recent years the use of public private partnerships (PPP) to fund the development of new buildings for the public sector has grown dramatically. These PPP projects involve a private developer paying for the building and then leasing them to the public sector over a long-term contract. In most of the early PPP projects the energy costs were typically transferred directly to the end-user and therefore there was no incentive to produce a low-energy building or to operate the building in an energy efficient manner. This issue has now been addressed in PPP guidelines, which recommend a sharing of the energy consumption risk (energy efficiency for PPP/PFI contracts website). Again, the role of the client in specifying a low energy building to start with is important. Public Private Partnerships can only be applied where the capital cost of the building exceeds £20 million.

LIMITED LIABILITY PARTNERSHIPS (LLPS)

One alternative approach to the structuring and financing of energy service projects (at least in the UK) would be to use the limited liability partnership (LLP) structure with both the client and the service provider taking a share in a special purpose vehicle (SPV) using an LLP. An LLP can be formed between individuals and/or organisations, which contribute either in cash or kind and gain a share in the benefits of the project(s). This structure could allow public bodies, or other organisations, to invest by contributing their utility/energy assets and share in the benefits from projects invested in by the service partner.

INVESTING IN NEW BUILDINGS

It has long been clear that incorporating energy efficiency features into new buildings or refurbishment projects is better than trying to add in or retro-fit projects after the building or facility has been completed. In order to do this successfully the client has to specify the requirement for energy efficiency, or more completely, sustainability, at the earliest stage of the project, even before concept designs are produced. The further down the design process you go before asking for energy efficiency and sustainability, the harder it is to incorporate it. For many years it was perceived that adding in energy efficiency to a building would add to capital costs but there is now a large body of evidence and experience that shows this is simply not true. Indeed, using the holistic approaches to engineering described above, capital costs may be reduced.

ESCO activity has been mainly confined to modifications to existing buildings and facilities. With the advent of tighter building regulations, property developers will be seeking new ways to achieve high levels of energy performance. At the same time emerging technologies such as BIPVs or building integrated wind power offer opportunities to turn buildings into generators of renewable power – power that has additional value in the market place. These trends suggest that ESCOs, particularly energy supplier-owned ESCOs, may find valuable opportunities by investing in energy systems within new buildings in return for long-term power supply and purchase agreements. There are some examples of ESCOs investing in new buildings in the USA (Eley Associates, 1997). The major problem of using any form of performance contracting for new buildings is, of course, deciding on the basis of comparison of performance between the new building and what would have been built without the ESCO intervention. This can be addressed by using a reference design and energy simulation.

SMALL AND MEDIUM SIZED ENTERPRISES (SMES)

Nearly all of the outsourced energy services activity to date has been in large enterprises. This is mainly due to the fact that outsourcing has considerable transaction costs and requires specific, bespoke technical development. These costs mean that it is not economic to apply the same models to SMEs. At the same time the SME sector accounts for a large proportion of total energy spend and often the potential to save is greater because SMEs do not have specialised technical and management resources to address energy issues. There is potential to develop new energy services within the SME sector that do not require bespoke technical development and which have low transaction costs. The issues of transaction cost and delivery mechanism are similar to those in the residential sector and there may be overlaps between the two sectors. Potential energy services for SMEs include aM&T provided through AMR, installation of proven 'black boxes' such as voltage reduction devices, lighting and heating controls, micro-generation through micro-turbines or micro-wind power installations. These could be paid for through supplements to long-term power and gas contracts that produce net savings to the customer. The diverse nature of the SME market, and the numbers of customers involved, suggest that utility-owned ESCOs should be well placed to address this market opportunity due to their existing customer base and their mass marketing and billing systems.

SUMMARY

A number of new technologies and engineering techniques are emerging that can have a significant impact on future energy consumption and energy systems. Because of their specialist knowledge, ESCOs should be able to foster the adoption and adaptation of these technologies and techniques before other non-specialist companies. There is also scope for innovation in financial and partnership structures. As well as pure energy services, a number of opportunities exist for expanding the market into the environmental arena.

Exploiting the potential for these new services will require innovation and an entrepreneurial approach on the part of both suppliers and customers.

Given the need to radically overhaul our energy systems and the emergence of both new technologies and new design techniques, it should be possible to combine these with innovative financing tools to produce very significant reductions in energy use and carbon emissions. This will require, however, a radical approach aimed at producing step-changes in efficiency and technology, rather than the incremental gains usually discussed.

Overall Summary

The energy issues facing all organisations, namely increasing energy prices and price volatility, climate change and increasing regulation, are not likely to diminish in importance in the foreseeable future. Managers of all organisations have to respond to these issues at a time when the technical, human and capital resources to do so have been reduced. The option of doing nothing is not viable for credible organisations and so the option is either to respond in a way that is focused on cost reduction, or in a way that gives strategic advantage.

As part of either response, outsourcing energy management is an important tool for organisations to reduce energy consumption and carbon emissions. This book has argued that outsourcing energy management has to be a carefully considered strategic decision rather than, as has often been the case, a tactical reaction. Any outsourcing project must also address all elements of energy and utility system cost, not just operations and maintenance or energy efficiency. A systems approach is needed. To date, much of the outsourcing activity has focused on one aspect of energy management, whether it be monitoring and targeting or identification and development of projects. Models under development as a result of supplier and customer partnerships are beginning to look at the total energy management picture and this approach will result in improved savings and reduced management of interfaces between different activities and suppliers. As these models are still developing this book, as well as hopefully being a useful guide to the outsourcing process, is a report on work in progress, work that is important for reducing costs, maintaining competitiveness and reducing environmental impact. There remains much work to do for everyone involved in this area.

Appendix 1: Strategic Analysis Pro Formas and Checklist

These pro formas are designed to guide completion of a strategic analysis. The tables shown, particularly the cost table, are simplified models.

STRATEGIC ANALYSIS

FUTURE ENERGY SCENARIOS

Power price forecasts (1, 3 and 5 year)
Gas price forecasts (1, 3 and 5 year)

CORE COMPETENCE

Is energy a core competence, meaning do we have a combination of skills and knowledge embedded in the organisation that results in the ability to execute utility operations to a world-class standard?

COST, RISK AND PERFORMANCE ANALYSIS PRO FORMA

CURRENT AND FUTURE COST ANALYSIS

A simplified current and future costs table is shown in Table A1.1. In most cases a more complex model linking energy use to future planned production will need to be assembled.

PERFORMANCE ANALYSIS

A sample performance analysis is shown in Table A1.2.

RISK ANALYSIS

A sample risk analysis table is shown in Table A1.3.

Table A1.1 Current and future energy cost analysis

	Year 1	Year 2	Year 3	Year 4	Year 5-10
Energy Costs					
Electricity Costs					
Gas Costs					
O&M Costs					
Labour cost					
Consumables					
Capital Cost					
Planned Item					
Planned Item					
Planned Item					
Total Cost					

Table A1.2 Performance analysis

Factor	Company performance	Industry benchmark (where applicable)
Overall energy efficiency (specific energy consumption – SEC)		
Utility labour productivity		
Quality of service - downtime - no. of failures of service		
Improvement in quality of service - over 1 year - over 3 years - over 5 years)		
Improvement in SEC (over 1, 3 & 5 years)		
Usefulness of energy information systems (ranking 1 to 5) - to senior management - to operational management - to all staff		
Outputs of each activity		
Costs at an activity level		

Table A1.3 Sample risk analysis

Risk	Probability (high/med/low)	Value at risk (£)	Avoid/Retain/ Reduce/ Transfer	Methods of avoidance, reduction or transfer
Commodity risks				
Commodity price				
Security of supply				
Quality of supply				
Conversion risks				
Efficiency				
Demand				
Correlation				
O&M				
Repair				
Replacement				
Project Risks				
Construction risk				
Performance risk				
Environmental				
CCL				
EU ETS				
IPPC				
Brand & image				
Environmental incident				
Perception/market				

CONSIDER SCOPE

A checklist for considering scope to be included in the outsourcing exercise is shown in Table A1.4.

CONSIDER PRICING OPTIONS

- fixed price
- cost plus

Table A1.4 Scope checklist

Activity	In scope of outsourcing
Establishment and operation of M&T system	
Provision of M&T information	
Motivation and awareness training	
Identification of projects	
Infrastructure replacement projects	
Energy efficiency projects	
Carbon management projects	
Development projects	
Financing of projects	
Project management of implementation	
Energy supply	
Market advice	
Energy supply risk management	
Invoice validation	
Operation and maintenance of utility systems	
Planned maintenance	
Unplanned maintenance	

- management fee
- hourly fee
- pay for performance
- gain sharing

Different elements of scope may require different pricing options.

CRITICALLY EXAMINE IN-HOUSE OPTION

- Are there any alternatives to outsourcing?
- Are there ways of restructuring energy management that would produce the desired results?
- If so what are their advantages/disadvantages?

Appendix 2: Request for Qualification – Energy Management and Project Development and Implementation

INTRODUCTION

The company is seeking to reduce energy consumption and carbon emissions as well as reduce total utility operating costs and risks. After a thorough review it has been decided to outsource all energy management functions to a suitably qualified service provider. This request for qualifications is designed to identify suitable service providers who have the requisite capabilities to form a long-term partnership with the company that will achieve our objectives.

SCOPE OF SERVICE

The scope of the service could include the following elements:

- supply of electricity and gas
- operations and maintenance of utility systems
- establishment and operation of a monitoring and targeting system
- provision of energy awareness training
- identification, development and implementation of utility systems upgrade, energy efficiency and carbon reduction projects
- financing of projects.

THE PROCESS

The expected process and timetable for seeking a partner is shown below:

- issue request for qualifications (RFQ) [date]
- evaluation of RFQ [date] to [date]

- shortlist service providers [date]
- issue request for proposals (RFP)
- evaluation of RFP [date] to [date]
- decision on preferred supplier [date]
- due diligence [date] to [date]
- negotiation of term sheet [date] to [date]
- negotiation of contracts [date] to [date]
- services commencement [date].

INFORMATION REQUIRED

Service providers are required to submit the following information:

Business details:

- company name
- date of incorporation
- registered office
- ultimate parent company
- years in business as an energy services company
- percentage of total revenue provided by energy services.

Office location from which this project will be managed:

Specific experience with similar projects including experience of:

- monitoring and targeting
- devising and running awareness campaigns
- identifying and developing energy efficiency projects
- identifying and developing renewable energy projects
- project management
- financing of energy efficiency and renewable energy projects
- carbon management

- supply of electricity and gas
- risk management.

Financial details:

- audited accounts for three years.

Legal actions:

- list any present or past legal actions against your company.

Health and safety policy and record:

- please provide a copy of the your health and safety policy
- please provide a record of all notifiable health and safety incidents over the last three years.

Environmental policy and record:

- please provide a copy of your environmental policy
- please provide a copy of your environmental report
- please provide a copy of any external certification relating to your environmental systems and record.

Quality systems:

- please provide a record of your quality process and describe your quality control systems
- please provide a copy of any certification for quality systems, for example ISO 9001.

Organisation details:

- CVs for all key personnel
- an organisational chart that describes your firm's project organisation
- a breakdown of numbers of staff according to the following levels:
  - managerial
  - professional

  - technical
  - manual.

Technical capabilities and details:

- a description of your technical and management capabilities in the areas listed under scope
- a description of your methodology for establishing baseline energy consumption for a similar facility
- a description of your approach to demonstrating savings achieved by projects
- a description of any relevant projects and the design approach used.

Sub-contractors:

- please provide list of any sub-contractors that you would use on this type of project
- please provide details of your systems for managing sub-contractors.

References:

- names and contact details for three reference sites (references will be contacted).

Appendix 3: Request for Proposals – Energy Management and Project Development and Implementation

BACKGROUND INFORMATION

SITE DESCRIPTION

A description of the site and data on the energy consumption and costs of the site, as well as the installed utility systems are shown in the Appendices[1].

Appendix 1. Site description

Appendix 2. Energy consumption and costs

Appendix 3. Description of utility systems

Appendix 4. Staff numbers by department

OBJECTIVES

The company's objectives are to reduce energy costs, carbon emissions and total utility operating costs and risks. It has been decided to seek a service provider who can form a long-term partnership with the company to achieve these objectives. The specific objectives of the outsourcing exercise are:

- to minimise energy consumption of the site through the provision of improved energy monitoring and targeting information and improving training and motivation of production staff ;
- to reduce energy consumption through the identification, development and implementation of energy efficiency and utility supply infrastructure projects;

1 The Appendices to the Request for Proposals referred to here will contain organisation-specific information and therefore are not reproduced in this section.

- to reduce carbon emissions;
- to reduce long-term utility operating costs;
- to reduce utility related risks.

The company is seeking a partner who can implement proposals that achieve these objectives at minimum total cost including energy costs, O&M costs and capital costs.

This request for proposals is not designed to be prescriptive in how these objectives will be achieved and tendering companies are encouraged to make specific proposals within their submission.

It is envisaged that a long-term agreement (five to fifteen years) will be signed between the parties. Service providers are asked to propose a suitable term.

SCOPE OF SERVICE

The scope of the service covers three areas:

- minimising energy consumption of the site through the provision of improved energy monitoring and targeting information and improving training and motivation of production staff;
- reducing long-term energy consumption and carbon emissions through the identification, development and implementation of energy efficiency and utility supply infrastructure projects;
- reducing overall carbon emissions through the identification, development and implementation of opportunities to use renewable and low-carbon electricity, either through the purchase of 'green' energy, local generation of renewable energy, and/or the purchase of carbon offsets.

Proposals to integrate these services with the supply of commodity (power and gas) will be considered, as will proposals to include the O&M of utility systems.

Minimising energy consumption through provision of improved energy monitoring and targeting information and improving training and motivation of production staff

The service provider will be expected to provide:

Steam and power consumption information for the following cost centres; [A], [B] and [C]. It is envisaged that this will require the installation of additional sub-metering. The following reports will be required for each cost centre and the entire site:

- daily energy use against target (power and gas)
- weekly energy use against target (power and gas)
- monthly energy use against target (power and gas)
- monthly energy spend (produced at month end plus 1 working day).

Reports will be made available over the internet/intranet and emailed to the following staff [list]. The service provider should demonstrate how data security will be maintained. Access to the existing site IT network may be available subject to suitable assurances on data and system security.

In addition, the service provider will be expected to provide appropriate energy awareness training to the numbers of staff shown in Appendix 4.

As well as initial training sessions it is envisaged that the service provider's energy manager will provide feedback and assistance to each cost centre based on the energy reports described above. Furthermore the service provider will provide refresher courses twice a year for the staff shown in Appendix 4. Service providers are expected to provide sample contents for energy awareness training as well as evidence of similar services provided elsewhere.

The service provider will be expected to provide an energy manager for the equivalent of two days a week for the duration of the contract. The energy manager shall attend the following meetings:

- monthly energy review meetings
- weekly production meetings
- site environmental meetings
- monthly contract review meeting
- range of other site meetings as required.

The energy manager shall ensure that energy performance data is provided in good time for discussion at these meetings.

Reducing energy consumption through the identification, development and implementation of energy efficiency and utility supply infrastructure projects

The service provider will identify projects designed to:

- reduce energy consumption
- improve the efficiency of energy conversion in utility supply equipment
- reduce water consumption
- reduce utility system operation and maintenance costs
- reduce carbon emissions.

The service provider will develop the project concepts sufficiently to assess their capital costs and savings within a +/–15 per cent error band. The supplier shall propose the optimum portfolio of projects to reduce the total long-term cost of utility provision (including energy and O&M).

The supplier shall propose their preferred method for funding and implementing these projects. These may include, but are not limited to:

- straight capital purchase
- performance contracting with client, service provider or external capital
- build, own, operate
- total utility system outsourcing.

The service provider shall also propose a programme of work, which will result in the proposed projects being implemented in a reasonable and practicable timescale taking into account factors such as design and engineering time, delivery times, and the need to maintain production schedules.

The selected service provider will then undertake due diligence to refine capital and savings estimates and work with the client to finalise the portfolio of projects to be implemented and the method of financing and implementation.

All projects must be designed and built using the appropriate engineering codes of practice and in a way that complies will all relevant legislation and regulation including, but not limited to: Building Regulations, L8 Legionella Guidelines, all relevant environmental regulations and health and safety guidelines.

All projects that reduce total operating costs (after taking into account all financing and other costs) will be considered for inclusion in the final agreement.

Service providers are expected to propose their methodology of project development including the use of measurement and verification (M&V) to aid in the sizing and design of projects.

Service providers should also specify their recommended M&V methodology to demonstrate post-project savings for each proposed project.

Reducing overall carbon emissions through the identification, development and implementation of opportunities to use renewable and low-carbon electricity, either through the purchase of 'green' energy, local generation of renewable energy, and/or the purchase of carbon offsets

The service provider will make proposals to reduce overall carbon emissions at least cost through either the purchase of green energy, local generation of renewable energy and/or the purchase of carbon offsets. The objective is to minimise carbon emissions at least total cost.

PRICING

Minimising energy consumption through provision of improved energy monitoring and targeting information and improving training and motivation of production staff

It is envisaged that this task could be charged for by a combination of:

- straight fee
- performance-related fee.

Reducing energy consumption through the identification, development and implementation of energy efficiency and utility supply infrastructure projects

It is expected that the service provider will price this task in a way that is related to the savings achieved (relative to the base year). The service provider will be expected to make proposals for the pricing mechanism for projects that include a performance element. The pricing of capital and other project costs will be on an open-books basis. The service provider will be expected to retain an ongoing interest in the long-term performance of the project.

> *Reducing overall carbon emissions through the identification, development and implementation of opportunities to use renewable and low-carbon electricity, either through the purchase of 'green' energy, local generation of renewable energy, and/or the purchase of carbon offsets*

The service provider is expected to make proposals for the pricing structure of these elements of the project. Any capital projects should be priced in a transparent manner as described in 3.2 above.

INFORMATION REQUIRED

Service providers will be expected to provide, at a minimum, the following as part of their submission in response to this RFP:

- name and address of the firm;
- statement of qualifications and experience (limited to five pages);
- name and contact details of person leading the response on behalf of the service provider;
- proposed monitoring and targeting software and hardware requirements;
- sample energy awareness training content;
- energy audit for facility [A];
- detailed project proposals for facility [A];
- optimum investment programme to achieve objectives at minimum total cost;
- proposed method(s) of financing projects;
- proposed approach to reducing overall carbon emissions including prices for green energy and carbon credits;
- CVs of all staff who will work on the project;
- an organisation chart for (a) the project and (b) the service provider;

- proposed methods for ensuring continuous improvement;
- names and contact details for three reference sites for which you have performed similar services. These references will be taken up during the evaluation period.

Do not submit advertising material or literature.

The response should be signed by an authorised officer of the firm.

ENQUIRIES

Enquiries regarding this RFP should be directed to:

[Contact]

[Address, contact details]

Any enquiries will be answered and the question and answer will be circulated to all bidders.

Service providers should not contact anyone other than the above named person.

RESPONSES

Responses must be received by [time] on [date].

Responses should be addressed to:

[Contact]

[Address, contact details]

Four hard copies and one electronic copy of the response are required. The hard copies should be clearly labelled: [label details].

GENERAL CONDITIONS

Companies responding to this RFP do so at their own cost.

[Other general conditions]

EVALUATION PROCESS

Following receipt of tenders, the submissions will be evaluated on the basis of the written submission and an oral presentation. The evaluation criteria are:

- business qualifications

- financial qualifications
- personnel qualifications and experience
- project management experience
- technical proposal
- experience of similar projects
- references.

The expected process and timetable is shown below:

- Evaluation [date] to [date]
- Decision on preferred supplier [date]
- Due diligence [date] to [date]
- Negotiation of term sheet [date] to [date]
- Negotiation of contracts [date] to [date]
- Services commencement [date].

References and Further Reading

REFERENCES

ACE (2005). *Asset Value Implications of Low Energy Offices. Phase 2 Report*. London: Association for the Conservation of Energy. Available online from: www.ukace.org/pubs/reportfo/2004% 20Asset%20 Value% 20Implications%20of%20Low%20Energy%20Offices%20-%20Phase%202%20report%20v1.01%20(print%20version).pdf

ALEKLETT, K. and CAMPELL, C. (2002). *ASPO Statistical Review of Oil and Gas, Proceedings of the 1st International Workshop on Oil Depletion*. Uppsala, Sweden: 23–25 May 2002. Available online from: www.peakoil.net/ASPOstatrew/ASPO-Stat-Rev.html

ALLIANZ GROUP and WWF (2005). *Climate Change & the Financial Sector: An Agenda for Action*. London: Allianz Group and WWF.

BERTOLDI, P. and REZESSY, S. (2005). *Definitions: Energy Service Companies in Europe Status Report 2005*. Luxembourg: European Commission, DG JRC, Institute for Environment and Sustainability, Renewable Energies Unit. EUR 21646 EN. Available online from: http://energyefficiency.jrc.cec.eu.int/pdf/ESCO%20report%20final%20revised%20v2.pdf

BP (2005). *BP Statistical Review of World Energy*. June.

BRIGGS, N. (2003). 'Power Quality – What is It?' *Engineering Talk*. Available online from: www.engineeringtalk.com/news/agk/agk164.html

COYNE, K.P., HALL, S.J.D. and CLIFFORD, P.G. (1997). Is Your Core Competence a Mirage? *McKinsey Quarterly* (1): 40–54. Available online from: http://instruction.bus.wsc.du/mcarpenter/readings/corecomp.html

DAVIDSSON, P. and BOMAN, N. (2000). 'A Multi-Agent System for Controlling Intelligent Buildings.' In Fourth International Conference on Multi-Agent Systems, Institute of Electrical and Electronic Engineers. New York.

DTI (2006). *Digest of UK Energy Statistics 2005*. London: Department of Trade and Industry. Available online from: www.dti.gov.uk/energy/statistics/publications/dukes/page19311.html

ELEY ASSOCIATES (1997). *Energy Performance Contracting for New Buildings*. Available online from: www.archenergy.com/library/general/perfcntr.doc

FISHER, R. and UREY, W. (2003). *Getting to Yes*. London: Random House Business Books.

FROOT, K. A., SCHARFSTENIN, D. S. AND STEIN, J. (2004). 'A Framework for Risk Management.' *Harvard Business Review*. November–December.

GIBSON, G., PRATT, D. and PATTERSON, J. (2001). 'Intelligent Software Agents for Scheduling of Distributed Generation.' Presented at POWER-GEN Conference, 12 December 2001. Available online from: www.aesc-inc.com/download/PG_Conference.pdf

GLA (2004). *Green Light to Green Power, the Mayor's Energy Strategy*. London: Greater London Authority.

GOLDMAN, C.A., OSBORN, J.G., HOPPER, N.C. and SINGER, T.E. (2002). *Market Trends in the U.S. ESCO Industry: Results from the NAESCO Database Project*. Berkeley, CA: Lawrence Berkeley National Laboratory. Available online from: www.naesco.org/industry/highlights/Market_Trends_in_the_ESCO_Industry_2002.pdf

GREAVER, M.F. (1999). *Strategic Outsourcing. A Structured Approach to Outsourcing Decisions and Initiatives*. New York: American Management Association.

GUINNESS ATKINSON FUNDS (2004). *The Future of Energy*. Investment Research Series. Woodland Hills, CA.

HANSEN, S.J. and BROWN, J.W. (2004). *Investment Grade Audit: Making Smart Energy Choices*. Lilbum, GA: The Fairmont Press.

HART, S.L. (2005). *Capitalism at the Crossroads. The Unlimited Business Opportunities in Solving the World's Most Difficult Problems*. Upper Saddle River, NJ: Wharton School Publishing.

HEYDT, G. (1991). *Electric Power Quality*. Stars in a Circle Publications.

HIRSCH, R.L., BEZDEK, R. and WENDLING, R. (2005). *Peaking of World Oil Production: Impacts, Mitigation and Risk Management* [online]. Washington, DC: US Department of Energy. Available online from: www.netl.doe.gov/publications/others/pdf/Oil_Peaking_NETL.pdf

HOUSE OF LORDS (2005). *The Economics of Climate Change*. Select Committee on Economic Affairs. 2nd Report of Session 2005–2006. London: The Stationery Office. Available online from: www.publications.parliament.uk/pa/ld200506/ldselect/ldeconaf/12/12i.pdf

LEGGETT, J. (2005). 'Half Gone: The Coming Global Energy Crisis, Its Conflation with Global Warming, and the Implications.' Presented at: 'Peak Oil UK: Entering the Age of Depletion'. Organised by Depletion Scotland. Edinburgh, 25 April 2005. Available online from: www.odac-info.org/documents/LeggettPresentation_000.pdf

MARTIN, P. (2005). 'Is Our Utility Industry in Denial?' *Energy in Buildings and Industry*, September 2005.

MCDONOUGH, W. and BRAUNGART, M. (2002). *Cradle to Cradle Remaking the Way We Make Things*. New York: North Point Press.

PATTERSON, W. (1998). *Electricity: Network Futures. Energy and Environmental Programme*: Royal Institute of International Affairs Transforming Electricity: Working Paper 4. London: Chatham House. Available online from: www.waltpatterson.org

RMI (2006). *Commercial/Industrial Energy Efficiency*. Available online from: www.rmi.org/sitepages/pid119.php

ROYAL SOCIETY. (2005). *Ocean Acidification Due to Increasing Atmospheric Carbon Dioxide*. Policy Document 12/05. London: Royal Society.

SHELL (2005). *Shell Global Scenarios to 2025. The Future Business Environment – Trends*. Washington, DC: Institute for International Economics.

SIMMONS, M. (2006). 'Energy in the 21st Century: A Rough Ride Ahead.' Presentation to Greater Kansas City Chamber of Commerce, Kansas City, MO, 21 February 2006. Available online from: www.simmonsco-intl.com/files/Kansas%20City%20Chamber%20of%20Commerce.pdf

SKREBOWSKI, C. (2005). 'Depletion – The Reality in Action.' Presented at: 'Peak Oil UK: Entering the Age of Depletion'. Organised by Depletion Scotland, Edinburgh, 25 April 2005. Available online from: www.odac-info.org/documents/chris_skrebowski.ppt

SORRELL, S. (2005). *The Contribution of Energy Service Contracting to a Low Carbon Economy*. Research Technical Report 37. Norwich: Tyndall Centre for Climate Change.

STERN (2006). *Stern Review. The Economics of Climate Change*. London: HM Treasury. Available online from: www.hmtreasury.gov.uk/independent_reviews/stern_review_economics_climate_change/stern_review_report.cfm

FURTHER READING

The following publications and web pages provide useful background reading.

PUBLICATIONS

ENERGY EFFICIENCY FOR PPP/PFI CONTRACTS FOR FURTHER AND HIGHER EDUCATION. *Building Research Establishment*. Good Practice Guide 321. Available online from: www.brepfi.com/Publications/Energy%20Efficiency%20in%20PPPPFI%20contracts%20for%20further%20%20higher%20education

JONES, A. (2003). *Woking: Local Sustainable Community Energy*. Presented at: 'Delivering the White Paper: Opportunities for Professionals and Local Communities to Work Together in New Ways'. Organised by the IEE Engineering for a Sustainable Future Professional Network, Doncaster, 21 May 2003. Available online from: www.iee.org/OnComms/PN/sustainability/event_210503.cfm

MCDOUGALL, G. and HINKS, J. (2000). 'Identifying Priority Issues in Facilities Management Benchmarking.' *Facilities* (18): issues 10/11/12.

WILLIAMS, B. (1994). 'Facilities Economics Incorporating Premises Audits.' Quoted in: BOTTOM, C. (2003). *Performance Measurement and Benchmarking in FM*. RICS CPD event Ireland. June 2003. Available online from: www.bwassoc.co.uk

WYMAN, R.A. (2004). *Disclosing and Managing Climate Change Risks in an Uncertain World*. White Paper. Los Angeles, CA: Latham Watkins. Available online from: www.lw.com/resource/Publications/_pdf/pub1212_1.pdf

ZWEIG, D. and JIANHAI, B. (2005). 'China's Global Hunt for Energy.' *Foreign Policy* (84):5.

Oil Peaking

CAVALLO, A.J. (2005). 'Oil: Caveat Empty.' *Bulletin of Atomic Scientists*, (61): 03.

CHIA, G. *There's No Fuel Like an Old Fuel*. Available online from: www.energybulletin.net/6819.html

HENDERSON, C. *Energy Wars and the Future of Planet Earth*. Available online from: www.opendemocracy.net/debates/article-6-32-873.jsp

KUNSTLER, J.H. (2005). *The Long Emergency*. New York: Grove/Atlantic Inc.

LEGGETT, J. *The Coming Crisis*. Available online from: www.earthscan.co.uk/news/article/mps/UAN/351/v/1/sp/

OIL DEPLETION CENTRE (2005). 'Peak Oil UK – Entering the Age of Depletion.' Conference 25 April 2005. Available online from: www.odac-info.org/PeakOilUKConferenceProceedings.htm

Peak Oil – 'Bottom Up' Approach Suggests Problems Begin in 2008. *Energy World* July/August 2005

SIMMONS, M.R. (2005). *Depletion and Peak Oil. A Serious Issue or an Over Exaggerated Fear?* 25 July 2005. Available online from: www.simmonsco-intl.com

SIMMONS, M.R. (2005). *Twilight in the Desert – The Coming Saudi Oil Shock and the World Economy*. Chichester: John Wiley & Sons.

Climate Change

RECHSTEINER, R. (2004). *Ten Steps to a Sustainable Energy Future*. 5 July 2004. Available online from: www.rechsteiner-basel.ch

SMALLEY, R.E. (2005). 'Nanotechnology and Our Energy Challenge.' Ohio Nanotechnology Summit, 2 March 2005.

SMALLEY, R.E. (2005). 'Future Global Prosperity: The Terrawatt Challenge.' *MRS Bulletin* (30) June.

SOCOLOW, R.H. (2005). 'Stabilisation Wedges: Mitigation Tools for the Next Half Century.' Keynote Speech on Technological Options at the Scientific Symposium on Stabilisation of Greenhouse Gases, 'Avoiding Dangerous Climate Change', Meteorological Office, Exeter, UK, 1–3 February 2005.

UK CLIMATE IMPACTS PROGRAMME. *A Changing Climate for Business: Business Planning for the Impacts of Climate Change*. Available online from: www.ukcip.org.uk

Regulation

The Climate Change Levy. Available online from: www.cclevy.com/ and from www.defra.gov.uk/environment/ccl/

EU Directive on the Energy Performance of Buildings. Available online from: www.managenergy.net/products/R210.htm

EU Directive on the Promotion of End Use Efficiency and Energy Services. Available online from: http://europa.eu.int/comm/energy/demand/legislation/end_use_en.htm

EU Emissions Trading Scheme. Available online from: http://europa.eu.int/comm/environment/climat/emission.htm

The Kyoto Protocol. Available online from: http://unfccc.int/2860.php

Integrated Pollution Prevention and Control (IPPC). Available online from: http://ec.europa.eu/comm/environment/ippc/ and from www.environment-agency.gov.uk/business/444217/444663/298441/?lang=_e

Operating & Financial Review. Available online from: www.dti.gov.uk/cld/financialreview.htm and from www.frc.org.uk/asb/press/pub0793.html%20 and from www.thecarbontrust.co.uk/carbontrust/reduce_emissions/OFR/MainOFR.htm

UK Building Regulations Part L. Available online from: www.odpm.gov.uk/index.asp?id–1130479

Why outsource?

HANSEN, S.J. and WEISMAN, J. (1998). *Performance Contracting: Expanding Horizons*. Lilburn, GA: The Fairmont Press.

HANSEN, S.J. (2002). *Manual for Intelligent Energy Services*. Lilburn, GA: The Fairmont Press.

Energy Management and Shareholder Value. Available online from: www.greenbiz.com

PETERSON, B.L. *Contracting for Energy Management Outsourcing*. Available online from: www.bpo-outsourcing-journal.com

The energy management process

FAWKES, S.D. (1987). 'Soft Systems Model of Energy Management and Checklists for Energy Managers.' *Applied Energy* (27): 3.

FAWKES, S.D. and JACQUES, J.K. (1988). 'Organising for Effective Energy Management'. In J. K. JACQUES, J-B. LESONID AND J-M. RUIZ (Eds) *Modern Applied Energy Conservation: New Directions in Energy Conservation Management*. Chichester: Halsted Press/John Wiley & Sons.

FAWKES, S.D., JACQUES, J.K. and WOOD, J. (1984). 'Agencies of Change in Energy Conservation – Applicable Models for Improving Energy Management Practices in the UK.' In *Energex '84: Energy Development: New Forms, Renewables, Conservation*. Toronto: Pergamon.

FAWKES. S.D. (2002). 'The Five Phases of Energy Management.' *Energy World* (295): 12–14. July.

ALLIANCE TO SAVE ENERGY (2003). *Strategic Industrial Energy Efficiency*. Available online from: www.ase.org/files/1012_file_SIEE.pdf

VESMA. V. (2002). *Monitoring and Targeting: Vilnis Vesma's Guide*. Available online from: http://vesma.com/mt/index.htm

The supply of outsourced energy solutions

DICKERMAN, R.N. (2005). 'Energy Outsourcing'. *Energy User News*, 24 April 2005. Available online from: www.energyusernews.com

OLDHAM, J., JAMES, P. and SHAW, B. (2003). *Delivering Resource Productivity: The Service Solution*. London: The Green Alliance. Available online from: www.worldenergy.org/wec-geis/publications/reports/liow/sustainability/sustianbility.asp

VINE, E.L. (2003). *An International Survey of the Energy Service Company (ESCO) Industry*. Berkeley, CA: Lawrence Berkeley National Laboratory.

Future trends for ESCOs

WEIZSACKER, E.U. (1998). *Factor Four: Doubling Wealth, Halving Resource Use*. London: Earthscan Publications.

INTERNET RESOURCES

As well as the specific internet sites listed above, the following sites have been found to be helpful. They represent a very small proportion of the information available on energy service contracts and projects on the internet.

www.eurocontract.net

Guide to energy performance contracting for Central and Eastern Europe

www.rwenpower.com

Leading UK energy and energy services provider

www.bsdglobal.com/tools/principles_factor.asp

Factor Four: Strategies and Tools

www.ipmvp.org

International Performance Measurement and Verification Protocol

www.mbdc.com

MBDC – Product and process design company dedicated to incorporating 'cradle to cradle' design into products and buildings

www.sustainable-development.gov.uk

UK government website on sustainable development

www.wupperinst.org/FactorFour

More on Factor Four ideas and best practice ideas

www.coloradoefficiencyguide.com/performance/default.htm

Guide to energy performance contracting

www.nef.org.uk/energyadvice/co2calculator.htm

CO2 emissions calculator

www.thecarbontrust.co.uk/carbontrust/reduce_emissions/OFR/CMKnowledgeCentre.htm

The Carbon Trust carbon management knowledge centre

www.energy-service-initiative.net

European Energy Services Initiative

www.peakoilclock.com/oiltheory.php

Peak oil website

Index

About the Author

Steven Fawkes studied Science of Resources at the University of Birmingham from 1977 to 1980, specialising in energy resources. From 1981 to 1984 he was at the University of Stirling where he completed one of the first PhDs in energy management. This work was practically-based and developed new models of energy management, and made contributions to the literature in both energy and in innovation. His career since then has been spent developing and implementing energy management programmes for organisations in the private and public sectors in several countries. As a consultant, he worked for many clients including local authorities with large and diverse property portfolios, in one case reducing energy consumption by 15 per cent across a portfolio of more than 1 000 buildings. He helped develop the Audit Commission's guide to energy management. During this period Steven was also a director of a company building low-energy houses and co-founded one of the first wind farm development companies in the UK.

From 1995 to 2000 he worked for the EU Phare programme, the European Bank for Reconstruction and Development, the United Nations Development Programme, the World Bank and USAID developing national energy efficiency programmes in Romania, several of which are now in operation. He also co-founded Energy Serv S.A., one of the first Energy Service companies in Romania. In 2001 he became General Manager of RWE Solutions UK where he developed and implemented innovative energy services contracts with leading companies including Sainsbury's and Diageo. In 2005 RWE Solutions became the energy services arm of RWE npower and Steven became Head of Energy Solutions, responsible for managing the existing business and developing new energy services incorporating energy supply, energy efficiency and distributed generation. In 2007 he joined a corporate finance company where he assists emerging energy companies to raise finance.

Photograph of the author reproduced with permission fom Portrait by Anthony Buckley and Constantine, London

Steven has published over 85 papers and articles, mostly on energy management and energy policy, including contributions to a text book on energy management published in 1988. As well as being a Chartered Engineer and Fellow of the Energy Institute, Steven has also completed a Diploma in Sustainble Management at the University of Cambridge.

His hobbies include reading, flying and promoting space tourism. Steven can be contacted at: stevenfawkes@aol.com. Further information can be obtained from: www.outsource-energy.co.uk